CLOUDS

JOHN A. DAY

THE BOOK OF CLOUDS

Published by Sterling Publishing Co., Inc.
387 Park Avenue South, New York, NY 10016
© 2006 by John A. Day
Distributed in Canada by Sterling Publishing
c/o Canadian Manda Group, 165 Dufferin Street
Toronto, Ontario, Canada M6K 3H6
Distributed in Great Britain by GMC Distribution Services
Castle Place, 166 High Street, Lewes, East Sussex, England BN7 1XU
Distributed in Australia by Capricorn Link (Australia) Pty. Ltd.
P. O. Box 704, Windsor, NSW 2756, Australia

10   9   8   7   6   5

Manufactured in China

All photographs © 2006 John A. Day except as listed on page 207

ISBN 13: 978-1-4027-2813-6
ISBN-10: 1-4027-2813-1

For information about custom editions, special sales, premium
and corporate purchases, please contact Sterling Special Sales
Department at 800-805-5489 or specialsales@sterlingpub.com.

Design: Richard J. Berenson
         BERENSON DESIGN & BOOKS, LLC
         New York, NY

THE BOOK OF

THE BOOK OF
CLOUDS

To Luke Howard (1772–1864)
my hero and inspiration,
who first gave clouds their Latin names
and became the Godfather of Clouds;
and to Jeffrey Howard Wright,
Luke's great, great, great, great grandson,
whom I found on the Internet
and who has become
my esteemed friend.

CONTENTS

Introduction

CLOUDS ARE SIMPLE ENOUGH, just a collection of ice crystals or water droplets visible to everyone. Yet they are a source of endless wonder. They appear in an infinite number of shapes and forms. Some are beautiful, some awe inspiring, and some, like the whirling funnel cloud, are terrifying. Clouds inspire artists, poets, songwriters. They have reminded astronauts, looking down from space, that Earth, a seemingly abstract orb, is a place of life and movement. Those great swirls of white—as they change shape, swell, evaporate into wisps, disappear and come back, glow with sunlight or darken with rain—are a constant reminder of how dynamic our planet is.

The Book of Clouds explains what clouds are, how they form, how they affect our planet, and how we can forecast weather just by observing them. In this book you'll find the names of clouds, learn how to identify them, and how to photograph them. Most of the pictures are from my own collection. I have been photographing clouds for much of my adult life, and have a Ph.D. in cloud physics. My knowledge of the physics of clouds and the science of meteorology has given me a great deal of satisfaction. Yet never does my soul feel so nourished as when I look up and really see the clouds—those elusive, captivating, ephemeral gifts of nature.

So, for the sheer enjoyment of it all, here are

Ten reasons to look up

I. Clouds and cloudscapes are the greatest free show on earth. It doesn't cost a penny to look up and feast your eyes on the view.

2. Clouds are never exactly the same. They come and go and take on different forms. While there are four basic cloud types (cumulus, stratus, cirrus, and nimbus), nature combines them to compose endless symphonies in the skies.

3. Many skies are simply beautiful to behold. There is no other way of saying it. The gradations of light and color in the late afternoon and very early morning hours are bouquets for the eye.

4. Clouds are a billboard of Coming Attractions. While it takes a skilled eye to interpret the messages on the billboard, there is a feeling of immense satisfaction when one's own forecast is verified.

5. Observing the sky at regular intervals makes one feel connected to nature.

6. Cloud watching promotes a global consciousness. Weather satellites bring large-scale images of cloud patterns into our homes. They help us realize that "our" clouds are connected to other clouds all around the world.

7. The earth is unique because of its vast amounts of water. Clouds are made of water and are a constant reminder of its importance.

8. Water is a miracle substance. Scientists have found that simple H_2O is anything but simple. Those H_2O molecules link together and bring us the glorious clouds above us. Without water, there would be no clouds.

9. Cloud watching is an antidote to boredom. Clouds are ever changing, ever evocative.

I0. Clouds are a magic show. Where do they come from, and where do they go? This is a mystery to the nonscientist, and an area of endless fascination.

	GROUP ONE Cumulus (Heap) Family	**GROUP TWO** Stratus (Layer) Family	**GROUP THREE** Heaps & Layers; Heaps in Layers	**GROUP FOUR** Precipitating Heaps
HIGH 25,000 feet and up	 **Cumulus Congestus** Page 34	 **Cirrostratus** Page 56	 **Cirrocumulus** Page 80	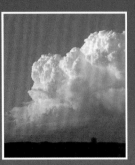 **Cumulonimbus** Page 82
MEDIUM 10,000–25,000 feet (Elevation range higher in summer, lower in winter)	 **Swelling Cumulus** Page 28	 **Altostratus** Page 52	 **Altocumulus** Page 72	 **Cumulonimbus** Page 82
	 **Fair-weather Cumulus** Page 22	 **Low Stratus** Page 48	 **Altostratocumulus** Page 68	 **Cumulonimbus** Page 82
LOW 0–10,000 feet	 **Fair-weather Cumulus** Page 22	 **Ground Fog** Page 40	 **Stratocumulus** Page 60	 **Cumulonimbus (base)** Page 82

GROUP FOUR Precipitating Layers	**GROUP FIVE** Optical	**GROUP FIVE** Optical	**GROUP SIX** Unusual	**GROUP SIX** Unusual

Cirrus
Page 94

Parhelia: Halo 22 degrees
Page 106

Aurora Borealis (Northern Lights)
Page 128

Contrails (Condensation Trails)
Page 160

Pileus
Page 154

Cirrus
Page 94

Parhelia: Sun Dog (Mock Sun)
Page 110

Corona
Page 120

Mammatus
Page 168

Lenticular "Flying Saucer"
Page 142

Nimbostratus
Page 92

Sun Pillar
Page 112

Rainbow (supernumerary)
Page 100

Virga
Page 170

Billow
Page 138

Nimbostratus
Page 92

Lightning
Page 182

Crepuscular Rays
Page 124

Cap cloud
Page 146

Sea Smoke
Page 158

How Clouds Form

IN 1968, when astronauts first headed for the moon, the most awe-inspiring sight they witnessed was the earth itself, hanging like an exquisite ornament against the utter blackness of space. The blues of the oceans and dark greens and browns of the continents were warm and inviting, a pleasing contrast to the monochromatic harshness of the lunar surface.

But more than anything, it was the ever-changing patterns of clouds that reminded the astronauts that their home was a place full of life. Those swirls of white, slowly shifting around the hemispheres, spoke of a dynamic planet, one of hot and cold, wet and dry, storm and calm.

It's unlikely that you'll ever get the chance to view the earth from 200,000 miles away, or even from a couple of hundred miles up in orbit. But an earthbound observer can still get a feel for the active nature of our planet, just by spending a few leisurely hours under the sky.

Start on a morning when the sky is nice and clear. Find a park or open field somewhere and bring a blanket and a picnic lunch along. Sometime after your post-lunch siesta, depending upon where you are and the time of year, there's a good chance that when you look up again, clouds will have appeared.

They might be bright fluffy ones, scattered across the skyscape like so many cotton balls. They might be thin wisps, chicken-scratchings in the high atmosphere. Or they might be dark, featureless expanses, rolling blankets obscuring the sun completely.

Whatever kind, those clouds are not part of some permanent collection, a cloud museum floating about the earth. No, those clouds were created—perhaps in the sky above you while you napped, but more likely somewhere else and then brought into your view by the wind. They formed where there appeared to be nothing, and at some point they will dissipate to seeming nothingness again.

It might seem like magic, this cloud creation, but in reality it is grounded in the earth—in the air and water, and in the fundamental physical principles that rule our world.

The Role of Water

A cloud is a simple thing, really—little more than water in liquid or frozen form. But it is a critical part of a process that is essential to life. That process is the hydrological cycle, which controls how water is distributed around the globe, where it nurtures crops and vegetation and slakes the thirst of the world's creatures, including its 6 billion people.

The water cycle is another simple thing: at its most basic it's a three-step process, involving evaporation, condensation, and precipitation. Water evaporates from oceans, rivers, and lakes, condenses into clouds in the atmosphere, and falls back down as rain and snow, eventually reentering the oceans, rivers, and lakes, where the cycle begins again.

Water itself is simple too. It's one of the most basic molecules known, just two atoms of hydrogen bonded to one of oxygen. But it has special characteristics that make it most suited to its role in the cycle that bears its name.

Water Is Everywhere

There is a lot of water on earth. Oceans cover about 70 percent of the globe's surface; and at the poles, huge additional amounts of water are stored as

Solid Water Floats

Water has another peculiar property: it is less dense as a solid than it is as a liquid. To put it another way, ice floats. Again, it is the molecular structure of water that is responsible. When liquid water molecules link up to form ice, they take up slightly more space. So given ice and water of equal volumes—a shoe box of each, say—there will be more molecules in the liquid box than in the frozen one, and the frozen one will weigh slightly less.

The implications of this for life on earth are enormous. For one thing, if ice were more dense than liquid water, it would sink to the bottom of lakes and ponds. Because it floats, it insulates the warmer water beneath it and protects marine life during the winter.

On a global level, because ice floats, it is exposed to sunlight, which aids in the melting and evaporation that is part of the water cycle.

Water Absorbs Heat

Liquid water is like an energy sponge. Because of the large difference in temperature between the melting point and the boiling point, water can soak up a lot of heat while remaining liquid. It's a huge sponge too—there is so much water on earth that it can soak up a great deal of heat from the sun and atmosphere without heating up much itself. As a result, oceans and lakes act like heat reservoirs, moderating the temperature fluctuations in air masses. One effect of this is that coastal climates tend to be less extreme than those inland.

The Role of Air

Clouds don't form in the oceans or ice caps, of course. They form in the atmosphere, the air.

Air fills every nook and cranny of the space around us. The tug of gravity pulls air molecules as close as possible to the earth. Air is thinner as elevations increase, and it is very thin at elevations of hundreds of miles. But no matter where it is, air is made up of the same gases. Two of these, nitrogen and oxygen, are plentiful, with nitrogen making up more than three quarters of air by volume, and oxygen, which is as critical to human survival as water, about one fifth. There are also small amounts of carbon dioxide, argon, hydrogen, methane, and other gases.

ice and snow. There is even some water in the atmosphere, in the form of vapor. It's a relatively small amount, but it plays a crucial role.

Water Exists in Three Forms

Water is unique in that, within the range of temperatures and pressures on earth, it occurs in all three states—solid, liquid, and gas.

The reason for this lies in water's molecular structure, in the way the two hydrogen and one oxygen atoms are arranged. Perhaps the best way to describe a water molecule is to say that it is a little out of electrical balance—there's a slight negative charge on the oxygen side and a slight positive charge on the hydrogen side. Since opposite charges attract, the hydrogen side of one water molecule forms a link—not a very strong one, but a link nonetheless—with the oxygen side of another.

Some of these links, called hydrogen bonds, must be broken in order to turn ice into water. All of them must be broken to turn water into vapor. It takes large amounts of energy to break the links. Surprisingly, the melting and boiling temperatures of water—32 degrees and 212 degrees Fahrenheit respectively—are much higher than those of molecules of similar organizational structure. If water were "normal," then at earth's temperatures it would only exist as a gas. There would be no oceans, no lakes, no rivers, and no rain—no water cycle at all. There would be no people, either, since the human body is largely made up of liquid water.

PRECIPITATION

TRANSPIRATION

WATER STORAGE IN
ICE AND SNOW

SURFACE
RUNOFF

GROUND
WATER
INFILTRATION

FRESHWATER
STORAGE

GROUND WATER
DISCHARGE

THE THREE STEPS OF THE WATER CYCLE:
Evaporation, Condensation, Precipitation

CONDENSATION
INTO CLOUDS

EVAPORATION

WATER STORAGE IN OCEANS

Also, like water, air has several properties that contribute to its role in the formation of clouds and the functioning of the water cycle.

Air Is Wet

There is always moisture in the atmosphere, caused by evaporation from the oceans, lakes, and rivers (and to a lesser extent by plant transpiration, the loss of water through leaves). But the amount can vary greatly. In the hot equatorial regions over the ocean, where lots of evaporation occurs, water vapor can make up as much as 7 percent of the volume of air. But in hot equatorial regions over land, such as the Sahara Desert, there is little evaporation. There is also little evaporation in the very cold polar regions, such as Antarctica.

Even in the same location, the amount of water vapor in the air can vary. Warm air can hold more water than cold air can, so in most places, summer tends to be more humid than winter. This difference in the amount of water vapor in warm air compared with cold is critical to cloud formation. When warm air with a lot of water becomes colder, some of the vapor condenses into liquid droplets.

To understand why this happens, one needs to realize that in warm air molecules move faster, while in cold air they move slower. When warm air cools, the molecules slow down. And as a water molecule in the air slows down, it has more of a chance to latch onto something else.

Often, what it latches onto is a bit of dirt.

Air Is Dirty

These days, we are all too familiar with smog and other types of air pollution. Obviously, air that can be seen as a brown haze on the horizon in an urban area is dirty. But even in pristine ecosystems, far from human activity, the atmosphere is not clean. Air everywhere is polluted with a variety of microscopic particles that can come from both natural and human sources.

Other than human activity, the main mechanism for introducing particles into the air is the friction of the wind against the earth's surface. When this occurs over the land, surfaces are slowly abraded and are broken down into small particles. Further breakdown in particle size comes about by the polishing effect of small particles rolling over each other.

Some of the dust from this polishing is fine enough to be carried aloft by the turbulent eddies in the wind, and it becomes more or less evenly distrib-

The eruption *of Mt. St. Helens in Washington on May 18, 1980, sent thousands of tons of ash into the upper atmosphere.*

uted throughout the lower atmosphere. Dust storms, such as those that plagued the central United States in the 1930s and those that plague Northern China and Mongolia today, are dramatic examples of this mechanism at work.

A second natural mechanism that operates infrequently but sometimes with massive effect is volcanic eruption. When the volcano Krakatau erupted in 1883 in Southeast Asia, for example, it produced a mantle of dust and ash that was noticed in the upper atmosphere around the northern half of the globe for several years afterward.

There are other natural mechanisms as well. Forest fires can pump huge amounts of smoke and ash into the atmosphere. The breaking of waves in the ocean and the bursting of air bubbles at the ocean's surface release salt particles into the air. Even the sun gets into the act, forming tiny particles of condensed gases called aerosols.

Aerosols, in fact, are a major source of atmospheric dirt. Some are formed by the action of sunlight on gases produced by automobiles and industry. But others are a product of solar radiation working on naturally occurring substances. There is evidence, for example, that certain compounds released by algae, when hit by sunlight, react to form aerosols. Similarly, terpenes, a family of compounds released by pine trees, are thought to be a source of sun-created aerosols.

Today we have a reasonable understanding of the various kinds of air-pollutant particles. In addition to knowing what they are and how they are produced, we know the range of concentrations in which they are found. Perhaps most important of all, we are coming to understand the role they play with respect to clouds: they are the seeds by which clouds form.

From Air to Cloud

We know that there is a lot of water in the world and that the air contains but a tiny fraction of it, in the form of vapor and cloud. We know that warm air can contain more water vapor than cold, and that when warm air with a lot of water vapor cools, some of the vapor condenses into liquid. And we know that the air contains a lot of microscopic particles of dust, minerals, and aerosols that the water can condense onto, forming droplets or, in the case of very cold air, ice crystals. It is these droplets and crystals that make up clouds.

But something is missing from this scenario. What makes warm air get cooler so that the water vapor in it condenses? In almost all cases, the answer is: the air rises.

Think of air pressure in terms of a stack of books. The heaviest weight is at the bottom of the stack and weight decreases as the stack goes up. Air pressure is like this. It is heaviest at its lowest level. On average, at sea level it is 14.7 pounds per square inch. It drops to half of this value at 18,000 feet, then decreases more slowly with increasing elevation.

Therefore, when a bubble of air rises, it finds itself surrounded by air of decreased pressure and expands so that the inner and outer pressure are the same. In expanding, the molecules of air within the bubble use some of their energy of motion—their kinetic energy.

Loss of energy means that the air gets cooler. The process is called adiabatic cooling, and the rate can be measured—air temperature drops by about 5.5 degrees Fahrenheit for every 1000 feet of altitude. (The process works in reverse as well—sinking air undergoes an increase in pressure and warms adiabatically.)

About 99 percent of clouds are formed as a result of such adiabatic cooling. (Some are formed at ground level, when warm air hits a cold surface.) But in order to cool in this way, of course, air must rise. Because of gravity's hold on it, air would normally move only horizontally—east or west, north or south. But it can ascend under certain conditions.

Hot Air Rises

Warmer air rises because it is less dense and thus more buoyant. A forest fire will do the trick, heating up the layers of air just above it. Sunlight striking land, which then radiates the energy as heat, can work as well. As the warm air rises, it is replaced by cooler, denser air. This circulation pattern is called a convection cell. The cooler air that replaced the warm air lowers the temperature of the rising column of air and raises the humidity. If humidity increases enough that the air is saturated, the condensation process begins. The clouds that result are cumulus-type clouds.

When Fronts Collide

Warm air can also rise abruptly when a mass of colder, denser air intrudes. The line of action is called a cold front. Because it is denser, the colder air hugs the ground; and as it moves in, it forces the warm air upward, as a snowplow would. Clouds at a cold front are cumulonimbus. Conversely, if a warm front moves into an area of retreating colder air, it creates a more gently sloping frontal surface and the clouds that form are very different, ranging from precipitating nimbo-

Clouds *tend to form, and most precipitation falls, on the windward sides of mountains.*

stratus to higher altostratus and still higher cirrostratus. Both effects are a major reason why plenty of clouds, often accompanied by thunder, lightning, and rain, pop up when a front moves in.

Hitting a Wall

Air also rises if it encounters a physical barrier, such as a range of hills or mountains. The air has to rise as it sweeps along the upward slope. The effect is called orographic lift, and it is the reason clouds tend to form, and most precipitation falls, on the windward sides of mountains, such as the western slopes of the Rockies.

No Place to Go

Finally, when several airstreams moving horizontally arrive at the same place, something has to give. There isn't enough space for all the air, and it cannot descend, because the ground is in the way. So some of the air has to rise. The effect is called convergence lifting.

A Cloud Grows

Once an air mass has risen and cools, it starts to humidify; that is, its relative humidity increases. Relative humidity is a comparison between the actual moisture in the air and the maximum amount possible at the given temperature. Cooling decreases the maximum possible moisture the air can contain. When the actual becomes the same as the maximum, the relative humidity is 100 percent. This is a condition of saturation, and it is the condition at which condensation begins. The cloud-formation process is now set in motion.

A cloud can be thought of as a miniature version of a galaxy, with the stars replaced by tiny water droplets on the order of 10 microns, or 10 twenty-five thousandths of an inch, in diameter. These droplets are far smaller than raindrops, but they are big enough to stay aloft in air currents and reflect light, giving clouds their customary white appearance.

Like galaxies, clouds are deceptive. A galaxy appears to be all stars; a cloud appears to be all droplets. And while there are billions of droplets in the average cloud, just as there are billions of stars in the average galaxy, there is actually far more empty space in both. If just doesn't look that way.

Those billions of droplets each grow from a seed, one of those aerosols or microscopic dust, salt, or mineral particles mentioned above. In cloud parlance, these particles have a special name: cloud condensation nuclei. They are very small, perhaps one tenth of a micron in diameter.

They are also hydrophilic, or water loving, with slight charges that can attract and hold on to water molecules. As the air cools and the water vapor molecules slow down, some of them are captured by the condensation nuclei. Soon the droplets are growing, as layer after layer of water molecules link (through hydrogen bonds) to the molecules already bonded to the condensation nuclei.

If the air is cold enough, the water will condense and freeze, forming ice crystals instead of droplets. But the process is the same—the crystals first form on condensation nuclei and grow as more and more water molecules link up.

In order for a cloud to form, this process has to occur billions of times over. Since the average diameter of a cloud droplet is roughly 100 times that of the seed on which it grows, the actual volume of the drop is a million times as big. So a droplet has to grow exponentially.

This growth can occur in several ways. Water molecules can be directly deposited on growing droplets, for example, but this is a very slow process. Growth occurs much more rapidly in some clouds in which there is a lot of vertical motion that causes droplets of different sizes to collide and unite into bigger ones.

Many clouds don't stop there. If the conditions are right, droplets will continue to collide and coalesce, eventually forming drops large enough to fall as rain. Again, phenomenal growth is required. A small raindrop—the kind you might encounter in a light drizzle—is about a millimeter, or one twenty-fifth of an inch, in diameter; so just one contains about a million droplets. An average-sized raindrop, on the other hand, might consist of more than 10 million droplets. And a large drop, the kind that falls when it's raining cats and dogs, might contain more than 100 million.

By then, the cloud is ready for a storm.

A PORTFOLIO OF

CLOUDS

The Cumulus Family

The term cumulus is Latin for heap or pile. Clouds in this family are usually detached from one another, with large areas of blue sky between them. While often flat on the bottom, their tops are shaped like domes or towers. Cumulus clouds that appear in a stable atmosphere are generally small benign clumps and are associated with fair weather. Unstable air can lead to cumulus congestus clouds. Very unstable air can bring massive, overpowering arrays of cloud that are made up of hundreds of cells thrusting high above the freezing level. These dramatic displays of cumulus spread out into an anvil shape and can become cumulonimbus clouds, which can bring thunderstorms.

Fair-weather Cumulus

Sometimes termed cumulus humilis, these widely spaced puffs of cloud resemble sheep grazing in a sky pasture. With flattish bottoms and rounded tops, they form in an ascending column of air caused by surface solar heating. Bases are generally from 2000 to 4000 feet. with cloud thickness from 1000 to 3000 feet. Favorite times for formation are summer afternoons in a benign atmosphere. In arid regions the bases and tops may be much higher. Cloudless spaces are regions of gently descending air.

A late spring day in Sussex, England. The sky is filled with puffy fair-weather cumulus clouds.

GROUP
Heap

NAME
Fair-weather Cumulus

BASE
2000–4000 ft.

TOP
4000–6000 ft.

AIR MASS STABILITY
Slightly unstable

BUOYANCY
Small positive

MOISTURE CONTENT
Low

TEMPERATURE
Above freezing

FRONTAL LIFT
None

PRECIPITATION TYPE
None

THE CUMULUS FAMILY
Fair-weather Cumulus

Summer afternoon clouds *form over an Oregon vineyard. A few especially active clouds push farther upward, to around 5000 feet. Bases are about 2000 feet. Some individual clouds are joining with neighbors to become stratocumuli.*

A late summer afternoon *in McMinnville, Oregon. Some cloud elements are dissipating, leaving only wisps of cloud (fractocumulus). Bases around 2500 feet. Tops 4000 to 5000 feet.*

A summer afternoon *in Wyoming. Cumulus images are reflected in the mirror surface of the lake. Wide variation is seen in the size of the individual clouds. Some are joining with others to become still larger cloud elements.*

THE CUMULUS FAMILY
Fair-weather Cumulus

Early summer afternoon *in the Willamette Valley in Oregon. Cloud elements in the foreground are starting to grow, while a few in the background have already moved to the swelling cumulus stage.*

As the earth's surface cools *at sunset, columns of heated air cool and the clouds begin to sink and dissipate.*

THE CUMULUS FAMILY
Swelling Cumulus

As the name implies, cumulus clouds in this group are developing towers and are swelling to greater heights as more intense surface heating gives them greater buoyancy. The upward forces come also from a more unstable atmosphere and other factors that initiate lift, such as the barriers of hills and mountains.

A road over hills *in southern Idaho. The image shows swelling cumulus clouds of moderate growth with bases at about 3000 feet and tops up to 12,000 feet.*

GROUP
Heap

NAME
Swelling
Cumulus

BASE
2000–4000 ft.

TOP
10,000–
15,000 ft.

AIR MASS
STABILITY
Moderately
unstable

BUOYANCY
Moderately
strong

MOISTURE
CONTENT
Moderate

TEMPERATURE
Above
freezing

FRONTAL LIFT
None

PRECIPITATION
TYPE
None

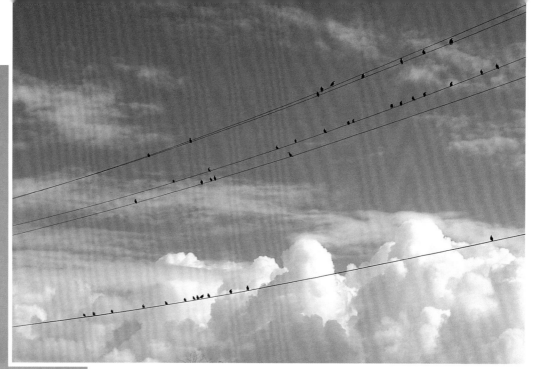

This field of cumulus towers *reaches about 15,000 feet. An appreciative aviary choir provides background music to the dramatic scene.*

Air rising over the slopes *of Kobuk Valley National Park in Alaska provides the upward thrust of this cumulus tower. Note several other towers in various stages of growth. The base is 13,000 feet and the top of the principal tower is about 22,000 feet.*

THE CUMULUS FAMILY
Swelling Cumulus

Low sun illuminates *the top towers of this multicelled cumulus mass. The top reaches perhaps 25,000 feet.*

Tens of separated towers *growing individually will join later to become a single massive cloud.*

THE CUMULUS FAMILY
Swelling Cumulus

A mountain of swelling cumulus *photographed in the light of the low sun. Some of its many towers reach about 20,000 feet.*

Swelling cumulus clouds *in different stages of growth, seen over hilly Sussex countryside, England. Bases are at about 2000 feet and the top of the largest elements is about 10,000 feet.*

THE CUMULUS FAMILY
Cumulus Congestus

The cumulus congestus is the last stage in development of the cumulus family prior to the cumulonimbus stage. Its multitude of cells is in all stages of development due to convective heating. The top of the cloud retains its hard-edged cauliflower appearance because the cloud elements have yet to reach the ice-crystal stage. Some light showers can fall from this type of cloud.

Three major towers *are made up of thousands of active convective cells.*

GROUP
Heap

NAME
Cumulus Congestus

BASE
4000–6000 ft.

TOP
15,000–25,000 ft.

AIR MASS STABILITY
Unstable

BUOYANCY
Strong positive

MOISTURE CONTENT
High

TEMPERATURE
Top reaching 32° F

FRONTAL LIFT
None

PRECIPITATION TYPE
Possible light showers

Cumulus Congestus

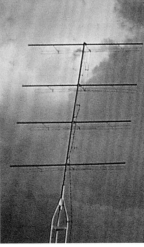

The unusual T-shaped cloud *above comes from strong local heating that causes atmospheric instability, which produces the initial tower. When they reach a certain elevation, the subsequent towers spread laterally as an atmospheric inversion level acts as a lid to upward growth.*

Inset, upper right:
A massive cloud mass, *extending upward to 35,000 feet. Such a cloud has strong up- and downdrafts and is on the verge of generating showers.*

This enormous cloud *consists of many thousands of convective cells in varying stages of development.*

Very active cumulus towers *are made up of thousands of convective cells in varying stages of development.*

A massive cumulus buildup *of many thousand convective cells is merging into the large cloud mass. The base is at 6000 feet, the top at 25,000 feet.*

A cloud resembling *a turkey, at sunset.*

THE CUMULUS FAMILY
Cumulus Congestus

The Stratus Family

Stratus, the name of one basic cloud type, is Latin for layer. This layered cloud is found at all levels, from ground fog to a veil of ice-crystal clouds called cirrostratus, formed at cold, high elevations. Stratus clouds are formless, sometimes thick, sometimes thin and wispy. Their color ranges through all stages of gray to pearl white, the color of cirrostratus clouds. There is no convective activity. Stratus forms when the atmosphere is stable.

Ground Fog

During the course of the night, the ground loses heat via radiation. Air touching the earth cools as a result. When the air temperature drops to the dew point, the air is saturated. Further cooling results in condensation of water vapor into tiny water droplets that are called, in the aggregate, ground fog. Think of ground fog as a layer cloud lying on the earth.

Radiation or ground fog, *which formed during an autumn night. Solar energy will warm the earth and start to thin out the fog.*

GROUP	Layer
NAME	Ground Fog
BASE	Ground
TOP	6000 ft.
AIRMASS STABILITY	Very stable
BUOYANCY	None
MOISTURE CONTENT	Low
TEMPERATURE	Wide range
FRONTAL LIFT	None
PRECIPITATION TYPE	Possible mist

Very thick ground fog, *with visibility reduced to 50 to100 feet.*

THE STRATUS FAMILY
Ground Fog

A grove of oak trees *is partially obscured by ground fog.*

A shallow layer *of ground fog forms in late afternoon in a meadow at the foothills of Oregon's Cascade Range.*

Low-lying stratus *over Lake Wakatipu at Queensland, New Zealand. Mountain peaks are visible through an opening in the clouds. New Zealand is known as Aetearoa in the Maori language, which means 'the Land of the Long White Cloud.'*

THE STRATUS FAMILY
Ground Fog

A flock of sea gulls *rests on a foggy Oregon beach. Visibility is only a few hundred feet. The fog will burn off later in the morning.*

Early morning mist is *rising from the relatively warm water surface.*
The shallow fog evaporates in the drier air above.
This kind of mist is sometimes referred to as sea smoke.

THE STRATUS FAMILY
Advection Fog

Thick advection fog forms
when warm moist air flows
over relatively cool water and
the air cools to its dew point.
If sunlight is not strong enough
to burn off the fog, the fog can
remain for many days. This
type of fog is called advection
fog. It often occurs in spring, as
the air begins to warm up,
as well as during winter thaws.

A fog bank enshrouds
*the shoreline of a bay
along the Pacific Coast.*

A river of fog *from the Pacific Ocean
pours under and over the Golden
Gate Bridge, making a ghostly scene
on San Francisco Bay.*

GROUP
Layer

NAME
Advection Fog

BASE
100 ft.

TOP
2000 ft.

AIR MASS
STABILITY
Very stable

BUOYANCY
None

MOISTURE
CONTENT
Low

TEMPERATURE
Wide range

FRONTAL LIFT
None

PRECIPITATION
TYPE
Mist

THE STRATUS FAMILY
Low Stratus

Low stratus clouds are formed when weak upward air currents lift a thin layer of air high enough to begin condensation. Light precipitation can fall from low stratus clouds, but generally the upward air currents are not strong enough to create significant rain. Wind conditions for stratus are calm or very light.

Near Bridgeport, California, *a layer of stratus nearly obscures a mountain basin. Note that the highway (the dark line, lower left) disappears under the cloud layer, reappearing in the far distance, some 15 miles away. The cloud has a very low base of 100 feet and a ceiling of 1500 feet.*

GROUP	
Layer	

NAME	
Low Stratus	

BASE	
1000–3000 ft.	

TOP	
2000–6000 ft.	

AIR MASS STABILITY	
Very stable	

BUOYANCY	
None	

MOISTURE CONTENT	
Low	

TEMPERATURE	
Wide range	

FRONTAL LIFT	
None	

PRECIPITATION TYPE	
Mist	

Wisps of stratus *hang low over the hills of the Salmon River Estuary on the central Oregon coast, allowing peeks at the surf line of the ocean.*

THE STRATUS FAMILY
Low Stratus

Low, wispy stratus *filled a river valley early in the morning but will dissipate in the heat of the day.*

THE STRATUS FAMILY
Altostratus

Altostratus occurs in the middle layer of the atmosphere. It is a pale layer cloud that produces overcast weather conditions. At times it can be translucent enough so that one can see the sun shining through it, as if through a piece of ground glass. Thickening altostratus generally means that rain is on its way.

The sky is entirely covered *by a layer of altostratus, varying in thickness and color.*

Midlevel clouds *thin below the sun, then thicken again near the horizon.*

THE STRATUS FAMILY
Altostratus

Sunlight emerging *from the dark cloud produces a bright path on the ocean waters. Note how the distortion of the sun's disk produces an upper knob. The coronal colors are due to diffraction of the sun's light as it passes through a cloud with uniform-sized cloud droplets.*

THE STRATUS FAMILY
Cirrostratus

These ice-crystal clouds are very high and semitransparent. They are easy to tell from the altostratus because the sun or moon can usually be seen through them, unlike the thicker altostratus, which only occasionally allows sun rays through. Cirrostratus clouds may cover the sky as a thin sheet, or they may have a fibrous look. They are also the clouds that sometimes produce a halo around the sun or moon. Cirrostratus is often a precursor to altostratus, then nimbostratus, which can bring rain or snow.

A sky filled with *ice crystals yields a fibrous appearance. Note the rather dim sun pillar.*

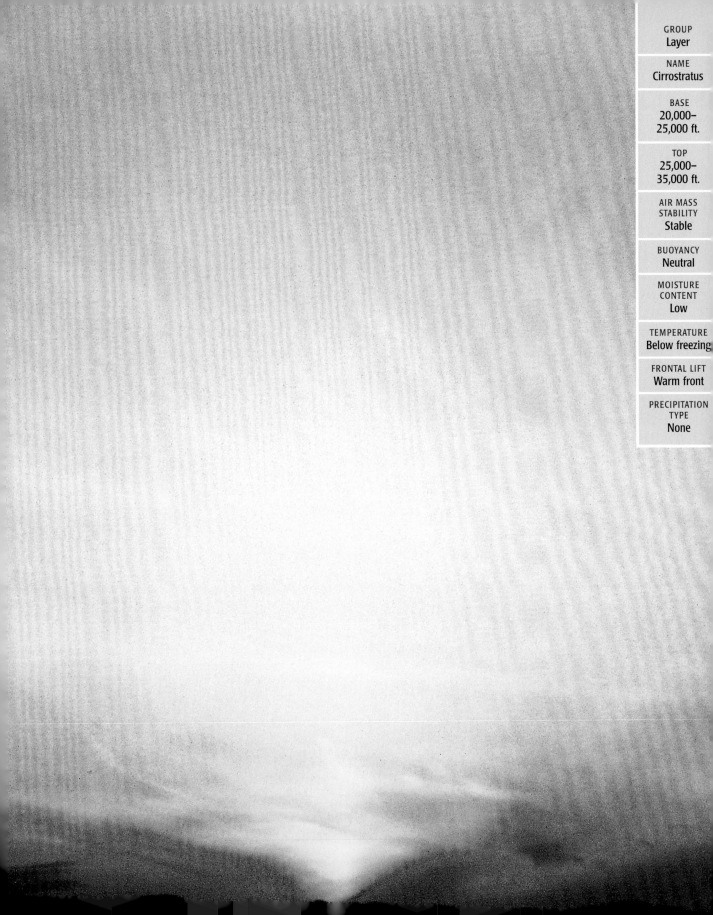

A sky of cirrostratus *at sunset provides a backdrop for the twin towers of St. Peter's Church and St. Paul's Church in Salzburg, Austria.*

THE STRATUS FAMILY
Cirrostratus

A visually thin layer *of ice-crystal cloud. Note the dim halo and the unusual coloration at the base of the sun pillar.*

Mixture of Cumulus and Stratus

The cumulus family consists of convective heaps of cloud formed in rising and cooling columns of air. The stratus family appears when there is no convection, for example when the air is stable and stratified and there is no vertical movement. An intermediate type of cloud forms when there is some convection with a slightly stable atmospheric layer. Hence we have stratocumulus, altocumulus, and cirrocumulus.

Stratocumulus

Stratocumulus is one of the most common clouds found in the low level of the atmosphere. It is a transitional form—between layered clouds in which there is no convection (stratus) and isolated cloud heaps (cumulus) formed when convection cells are present. Stratocumulus clouds are marked by irregularly shaped globules of cloud separated by clear or thin space. The clear space between clouds indicates sinking air that is both warm and dry. The area occupied by clouds is indicative of rising, cooling, and humidifying air.

The New Mexico skies are notable for their special photographic light. This scene of a very old adobe structure near Abiquiu shows a sky predominantly full of gently swelling cumulus joined by stratocumulus fragments.

GROUP	Heaps and/or layers
NAME	Stratocumulus
BASE	2000–5000 ft
TOP	4000–7000 ft
AIR MASS STABILITY	Slight instability
BUOYANCY	Small positive
MOISTURE CONTENT	Moderate
TEMPERATURE	Wide range
FRONTAL LIFT	None
PRECIPITATION TYPE	None

Inset, upper right:
This layered sky of *stratocumulus clouds shows strong evidence of convection activity. Dark areas correspond to ascending air, clear areas to descending air.*

MIXTURE OF CUMULUS AND STRATUS
Stratocumulus

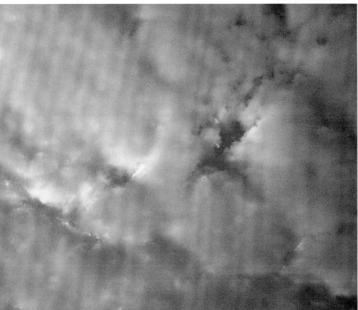

This kind of cloud *formation forms what is called a buttermilk sky, with large globules of convection cloud separated by clear space.*

A wide layer *of gray cloud, showing cumuliform lumpiness, covers the sky over a Wyoming grassland.*

MIXTURE OF CUMULUS AND STRATUS
Stratocumulus

A mix of cumulus and stratocumulus *floating over the Grand Canyon.*
Note the wide range of cumuliform developments—from small cloud
fragments to a few swelling cumuli—seen clearly in the center of the photo.

A disturbed, chaotic sky, *filled with misshapen cumuli and strati.*

MIXTURE OF CUMULUS AND STRATUS
Stratocumulus

Stratocumulus *organized in parallel rows.*

Here a dramatic sky *shows a mix of heaps and layers of cloud.*

MIXTURE OF CUMULUS AND STRATUS
Altostratocumulus

Altostratocumulus clouds are a mixture of layer and heap clouds that form in the middle atmosphere. Some are altostratus, others altocumulus. They are lower than cirrostratus clouds and are thicker. Unlike cirrostratus, these clouds do not permit the sun or moon to shine through, and they do not produce halos. Altocumulus clouds have more of the "heap" look, while altostratus are more layered.

Sunset *at Mount San Gorgonio near Palm Springs, California.*

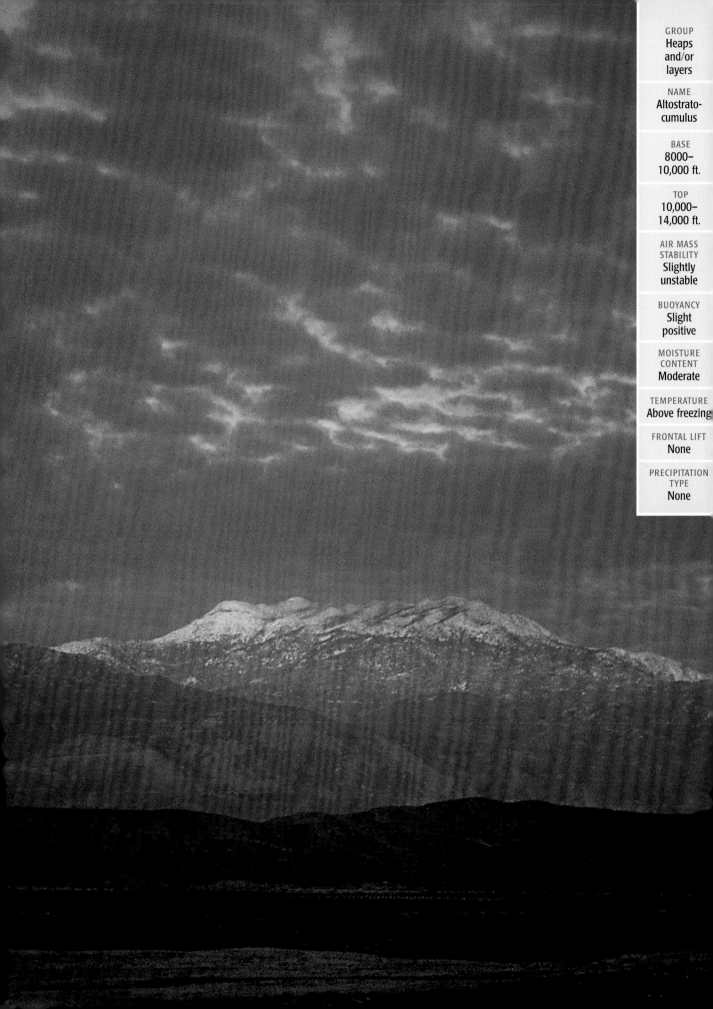

A clock tower *juts majestically upward toward the base of a buttermilk sky of altostratocumulus. Breaks between cloud globules are due to descending air.*

MIXTURE OF CUMULUS AND STRATUS
Altostratocumulus

An uneven sky *with a great deal of movement over Worm's Head in Wales. Here the clouds are at about 6000 feet. The air is oscillating, with clouds in the up region and clear space in the down region.*

MIXTURE OF CUMULUS AND STRATUS
Altocumulus

This is a midlevel layered cloud dimpled with convection cells that indicate a condition of instability and overturning in the atmosphere. These clouds are sometimes arranged in rows or small ripples.

A classic altocumulus sky *clearly shows layer and heap characteristics.*

A sheet of altocumulus *photographed in the last rays of the setting sun.*

GROUP
**Heaps
and/or
layers**

NAME
Altocumulus

BASE
**12,000–
16,000 ft.**

TOP
**14,000–
18,000 ft.**

AIR MASS
STABILITY
**Moderate
instability**

BUOYANCY
Positive

MOISTURE
CONTENT
Moderate

TEMPERATURE
Above freezing

FRONTAL LIFT
None

PRECIPITATION
TYPE
None

High altocumulus
layer approaching the
cirrocumulus level.
Note how the clouds'
billows look like fringes
at the edge of a rug.

MIXTURE OF
CUMULUS
AND STRATUS
Altocumulus

An altocumulus sky
after sundown.

A jiggly Jell-O sky with active convection in the layered cloud. Agitated up and down motion is evident even in this photo.

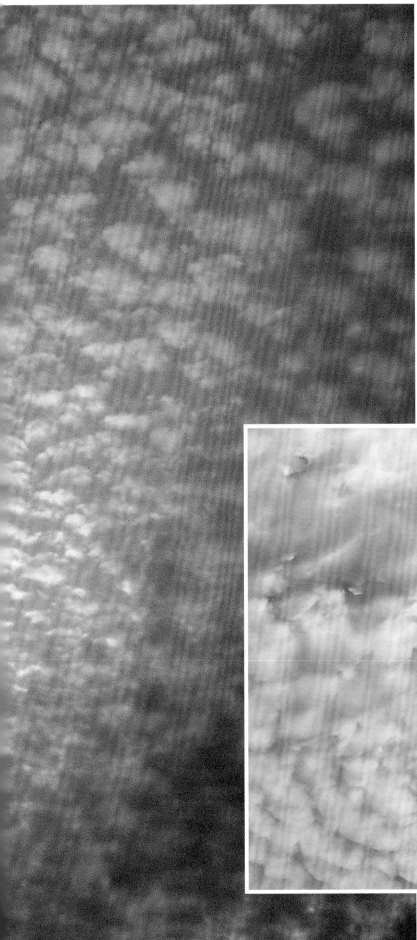

Many small convective *elements can be observed in this high cloud layer, with elevation about 20,000 feet.*

MIXTURE OF CUMULUS AND STRATUS
Altocumulus

Relatively low *and thick elements of altocumulus.*

MIXTURE OF CUMULUS AND STRATUS
Altocumulus

The cloud layer opens *to allow sunlight to penetrate. Note the billow organization on the left. This sky is sometimes referred to as a mackerel sky, as the cloud elements resemble fish scales.*

Convective elements cluster in lower left of the picture. They become smaller and more separated as one's eye scans up.

A thick sheet of altocumulus photographed in the light of the low sun.

MIXTURE OF CUMULUS AND STRATUS

Cirrocumulus

Cirrocumulus clouds are thin white patches or layers of cloud, composed of very small elements in the form of grains or ripples. They are more or less regularly arranged. Each cloud globule is the upward stem of a convection cell.

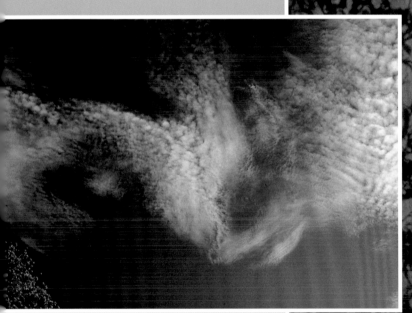

A wide variety *of cloud elements caused this odd-looking sky. The clouds' elevations reach 25,000 feet.*

This sky is filled *with tiny globules of convective cells.*

GROUP	Heaps and/or layers
NAME	Cirrocumulus
BASE	20,000–25,000 ft.
TOP	23,000–27,000 ft.
AIR MASS STABILITY	Slight instability
BUOYANCY	Slight positive
MOISTURE CONTENT	Low
TEMPERATURE	Near or below freezing
FRONTAL LIFT	None
PRECIPITATION TYPE	None

Precipitating Clouds

Precipitation falls from heap and layer clouds. The name for the heap precipitator is cumulonimbus. These clouds produce localized heavy showers. The name for the layer precipitator is nimbostratus. These produce widespread, continuous rain. Light showers sometimes fall from well-developed cumulus congestus clouds. Very light mist or drizzle sometimes falls out of low but thick stratus clouds.

Cumulonimbus

Cumulonimbus is the "granddaddy" of the cumulus family and can provide theatrical sky shows. As the growing cloud pushes above the freezing level, the top takes on a striated appearance due to the presence of ice crystals. Showers, often heavy, fall from the base. In strong updraft conditions, there may be showers of hailstones ranging from pea-size ice balls to golf-ball size and, on rare occasions, even to baseball size. The cumulonimbus often is associated with displays of lightning, and when the clouds are extremely active, tornado funnels may drop from the base.

This massive *cumulonimbus is fairly young, as the tops are just beginning to show ice-crystal striation. The cloud mass contains many thousands of convection cells, each one releasing buoyant energy.*

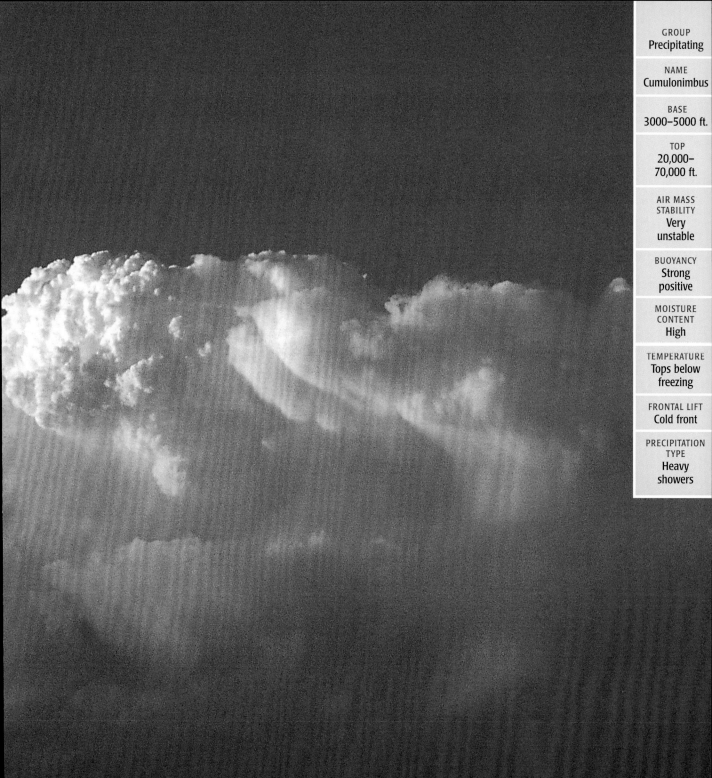

GROUP
Precipitating

NAME
Cumulonimbus

BASE
3000–5000 ft.

TOP
**20,000–
70,000 ft.**

AIR MASS
STABILITY
**Very
unstable**

BUOYANCY
**Strong
positive**

MOISTURE
CONTENT
High

TEMPERATURE
**Tops below
freezing**

FRONTAL LIFT
Cold front

PRECIPITATION
TYPE
**Heavy
showers**

A massive cumulonimbus sky,
*photographed at Boulder, Colorado.
Here the anvil top occurs when
the upward thrusting cloud tries
to penetrate the stratospheric lid
and spreads laterally. The cloud top
is estimated at 70,000 feet.*

PRECIPITATING CLOUDS
Cumulonimbus

Two towers erupt, *with cloud
tops showing signs of the ice
crystal formation process.
What was once a hard
"cauliflower" edge now has
a stringy appearance.*

The anvil that tops this cumulonimbus *tower is being blown off to the right by strong winds in the high troposphere.*

PRECIPITATING CLOUDS
Cumulonimbus

This cumulonimbus *shows a well-developed anvil top that resulted from the cloud's encounter with the higher, stable stratospheric level. This level diverts upward currents to a lateral motion causing the anvil shape.*

A dramatic break *in cumulonimbus storm clouds occurs in Amarillo, Texas. With the break comes the promise of better weather.*

A very large cumulonimbus *is photographed in low sunlight in Cheyenne, Wyoming. The top of the cloud flattens as it reaches the base of the stratosphere at 65,000 feet.*

PRECIPITATING CLOUDS
Cumulonimbus

The top of the cumulonimbus *mass to the right is completely transformed to ice crystals. Note the resulting stringy appearance.*

A lone tree at Amity, Oregon. *A magnificent cumulonimbus in the background spreads out and is in the early stages of disintegration.*

PRECIPITATING CLOUDS
Cumulonimbus

The pilot of a jetliner *skirts an immense cumulonimbus at 35,000 feet, northwest of Atlanta, Georgia. The cloud's top is about 60,000 feet.*

PRECIPITATING CLOUDS
Nimbostratus

Nimbostratus is the middle level (10,000 to 25,000 feet) cloud from which continuous rain or snow falls. The least photogenic of all cloud forms, it results from stable air ascending up and over the gentle slope of a warm front. In sharp contrast to the localized showers and squalls that fall from the base of cumulonimbus clouds, nimbostratus precipitation is widespread and long lasting.

A snowy scene *over the Syskiyou Mountains of southern Oregon. Snow has fallen from nimbostratus clouds.*

GROUP	Precipitating
NAME	Nimbostratus
BASE	6,000–10,000 ft.
TOP	15,000–18,000 ft.
AIR MASS STABILITY	Very unstable
BUOYANCY	Neutral
MOISTURE CONTENT	Moderate to high
TEMPERATURE	Slightly above freezing
FRONTAL LIFT	Warm front
PRECIPITATION TYPE	Moderate to heavy; continuous

PRECIPITATING CLOUDS
Cirrus

Luke Howard, who named the cloud types, described an upper cloud as resembling a lock of a child's hair. He gave it the Latin name for curl, cirrus. Cirrus is perhaps the most photogenic of all clouds. Delicate is the most descriptive term for the wispy appearance of the ice streamers that comprise this cloud. Generally white, cirrus sometimes has a silky sheen.

One may wonder why cirrus is grouped as a precipitating cloud. This is a valid question. In a four group classification scheme, it simply fits better here, though not perfectly, than it does under Groups 1, 2, or 3. Streamers of ice crystals fall—precipitate—and usually evaporate before reaching the ground. Thus the technical connection.

Streamers *of thick cirrus stretch eastward over the Front Range of the Rockies in central Colorado.*

GROUP	Precipitating
NAME	Cirrus
BASE	20.000–30,000 ft.
TOP	30,000–40,000 ft.
AIR MASS STABILITY	Slightly unstable
BUOYANCY	Small
MOISTURE CONTENT	Low
TEMPERATURE	Below freezing
FRONTAL LIFT	None
PRECIPITATION TYPE	Showers

The sky *over this lovely beach grass on the Oregon coast is filled with jet-stream cirrus.*

PRECIPITATING CLOUDS
Cirrus

This sky is covered with varied cirrus, *thick masses and thin filaments. In the foreground is a very old pioneer cabin, standing alone in the wilds of western Utah.*

Cirrus clouds scatter *through the evening sky in Eton, England.*
The low sun is reflected on the surface of the river Thames.

Cirrus

A dramatic outbreak *of cirrus filaments, both thick and thin, over McMinnville, Oregon.*

9:18 a.m. in Portsmouth, *New Hampshire. The sky is overflowing with hooked cirrus.*

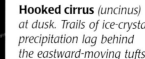

Hooked cirrus *(uncinus) at dusk. Trails of ice-crystal precipitation lag behind the eastward-moving tufts.*

Optical Effects

Under the proper atmospheric conditions, spectacular optical effects can be seen in the sky. Making this display possible is the very nature of sunlight. Sunlight is a white light made up of a composite of a spectrum of colors. When sunlight interacts with water droplets or ice crystals, the white light separates into its component colors. This can produce rainbows and parhelia—bright spots often tinged with color. Light may also be reflected and refracted; it can produce interference waves or be diffracted. All of these actions produce fascinating optical effects.

Rainbow

A rainbow is one of nature's greatest gifts. Rainbows are groups of concentric arcs in colors that range from violet to red. They are formed when drops of water from a rain shower are lit by light from either the sun or the moon. As the reflection and refraction of light interacts with the drops of water, a magnificent display of color results. The rainbow is usually a 180-degree arc, sometimes less. In a special confluence of events—at noontime and when viewed from above, as from a plane, a 360-degree ring of colors can be seen.

The wonderful sight of a rainbow.
Jonathan Livingston Seagull was in the right place at the right time.

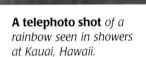

A telephoto shot *of a rainbow seen in showers at Kauai, Hawaii.*

OPTICAL EFFECTS
Rainbow

An especially *brilliant rainbow, photographed in Oregon's Willamette Valley. This is a supernumerary rainbow—note the faint secondary bow.*

Rainbow

A vivid primary bow *with dim secondary bow, seen at dusk in Pelican, Alaska, a small fishing village on the state's southeast peninsula. Note how the still waters of the harbor reflect the colors of the rainbow.*

A supernumerary rainbow,
seen near Salem, Oregon.

This telephoto shot
*reveals a vividly
colored supernumerary
rainbow. Note the
interior repetitive bows.*

OPTICAL EFFECTS
Parhelia: HALO

Parhelia produce dramatic optical effects. There are a number of subtypes of parhelia. Shown in this section are halos, sun dogs, and sun pillars. Parhelia are seen in many forms, including circles around the sun or the moon, bright spots on either side of the sun or moon, colored rings, arcs, and pillars. They result from the reflection of sunlight from the faces of ice crystals and/or the refraction of light as it passes through the crystals in the atmosphere.

A 22-degree small halo is intersected by a jet's narrow contrail at 35,000 feet.

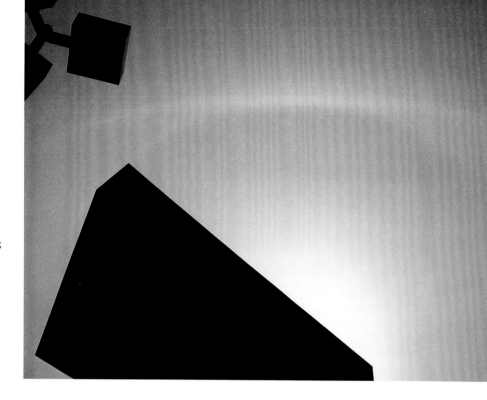

A small 22-degree halo is formed when sunlight shines through cirrostratus clouds. The sign is used to shield the camera lens from direct sunlight.

A 22-degree small halo. *The sun's orb is intentionally obscured by a flag for photographic purposes.*

OPTICAL EFFECTS
Parhelia: HALO

A 22-degree small halo *photographed at sunset. Cirrostratus thins at the top of the arc.*

OPTICAL EFFECTS
Parhelia: SUN DOG

Sun dogs, or mock suns as they are sometimes called, are two bright spots of colors at 22 degrees, usually found on either side of the sun at the same elevation. Sun dogs are produced by the refraction of light passing through ice crystals.

A single sun dog *shines brightly in the extended ice crystal plane from a mother cumulonimbus cloud to the left.*

OPTICAL EFFECTS
Parhelia: SUN PILLAR

Sun pillars are usually seen at sunrise or sunset, as the angle of a low sun creates these magnificent effects. The shafts or pillars of bright white light stretch from the ground to the sky above. They are caused by light reflecting from horizontal surfaces of hexagonal ice plates.

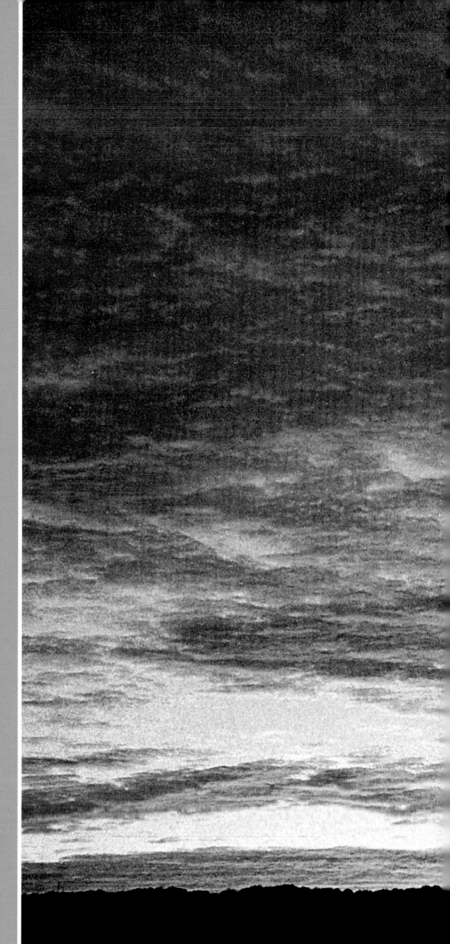

A sun pillar *of white light with a small halo, at sunset.*

Sun Pillar

A 22-degree *small halo is seen at dusk with a sun pillar.*

The same view *as above, later in the evening.*

OPTICAL EFFECTS
Irisation

Iridescence is seen in clouds when drops of water that are the same size diffract sunlight. Pastel colors, sometimes similar to the hues found in mother-of-pearl, appear at the edges of thin clouds, usually altocumulus.

Note how iridescence *produces pastel colors in the thinning edge of a sheet of altostratus.*

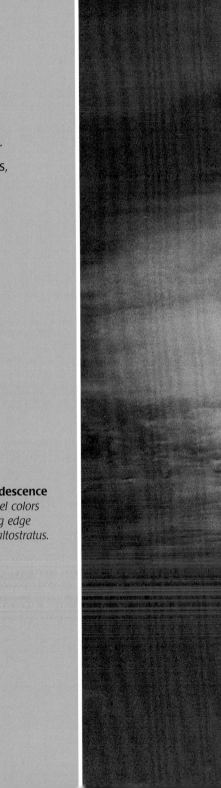

OPTICAL EFFECTS
Irisation

Thin edges *of thick altostratus clouds reveal pastel irisation colors.*

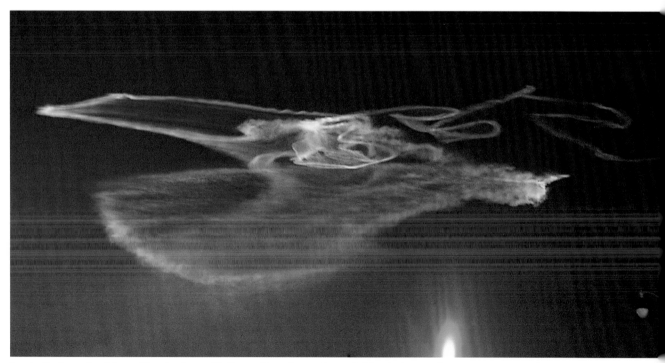

A missile launched from Vandenberg Air Force Base *in southern California bound for the Western Pacific. Here it creates strange-appearing stratospheric clouds, gleaming with iridescence. The photo was taken at approximately 8:00 p.m.*

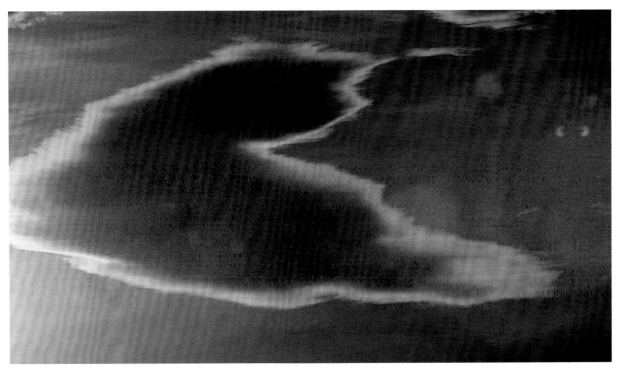

Looking like an aerial manta ray, *this cloud displays vibrant iridescence at its edges.*

A brilliant sun dog image *appears in a cumulonimbus anvil.*

Corona

Coronas are colored rings of small diameter centered about the sun or moon. They are caused by the diffraction of light as it passes through a cloud of small droplets, usually altocumulus. The size of the droplets must be uniform for the phenomenon to occur. The radius of the rings increases as the size of the droplets decreases. The inner color of the corona is blue or violet.

When sunlight interacts *with cloud droplets of an altostratus, the diffraction produces coronal colors around the sun's disk.*

Corona

A dramatic corona *colors the orb of a full moon.*

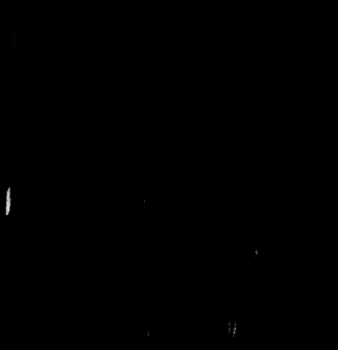

In the evening hours, *a corona around the moon is dramatically visible over England's Windsor Castle.*

OPTICAL EFFECTS
Crepuscular Rays

Crepuscular rays are beams of sunlight that seem to radiate from the sun. Since they radiate from one source toward the viewer, they are actually parallel to each other; but through a trick of perspective, they appear as divergent rays. They form as sunlight passes through gaps in stratocumuliform clouds and is brightly scattered in the lower atmosphere, giving the dramatic ray appearance.

Crepuscular rays
caught at twilight in the Virgin Islands.

OPTICAL EFFECTS
Anticrepuscular Rays

Anticrepuscular rays converge on the antisolar point, which is the point on the sky that is opposite the sun. While these rays are not rare, they are difficult to spot. One way to see them is after spotting a crepuscular ray. Turn your back to the sun and you should be able to find anticrepuscular rays. As with the crepuscular rays, these are parallel shafts of light that, through a trick of perspective, seem to converge in the distance.

These difficult-to-spot *rays are seen here on the edge of a small town in the northwest.*

OPTICAL EFFECTS
Aurora Borealis

The aurora borealis, or northern lights, almost defies description. It is ephemeral, a shimmering veil, mysterious and ghostly. It owes its existence, as does all of our weather, to the energy of the sun. An enormous amount of radiant energy streams from the sun, sometimes in the form of flares that shoot toward the earth in a phenomenon known as solar wind. Carried in this solar wind are charged particles. When the particles reach the outer edge of the earth's atmosphere, they interact with the earth's magnetic field. The magnetic field disperses the particles to the polar regions. There they react with gases in the earth's upper atmosphere, and brilliant flashes of light occur. These flashes of light are called the aurora borealis, or northern lights, in the Northern Hemisphere. In the Southern Hemisphere, they are called aurora australis. In the United States, the aurora borealis can sometimes be seen in the northern part of the country, very infrequently in the south.

The magnificent colors *of the aurora borealis are captured high above a stand of pines in northern Pennsylvania.*

OPTICAL EFFECTS
Glory

The glory is a complex optical phenomenon. It is seen when an object's shadow is projected on a cloud. Due to diffraction, colored rings are produced around the perimeter of the shadow.

A glory optical *effect is seen around a man's shadow. The photo was taken in Colorado.*

OPTICAL EFFECTS
Green Flash

This well-named phenomenon occurs when a momentary flash of brilliant emerald-green light appears just above the sun as it drops below the horizon. Green flashes are best seen over the sea, under cloudless conditions, in a relatively clear atmosphere.

These remarkable *three-stage shots of the ephemeral green flash phenomenon were taken at sunset off the coast of Cayucos, California.*

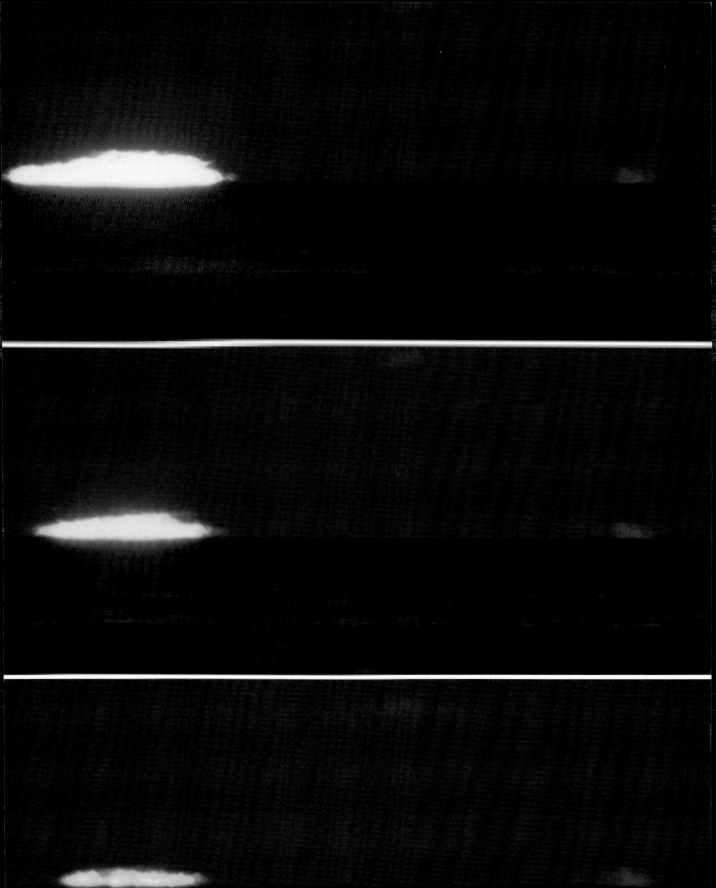

OPTICAL EFFECTS
Sunrise/Sunset

The glorious colors often seen in clouds at sunrise and sunset are due to the scattering of different wavelengths of sunlight. The scattering is caused by dust particles and water vapor in the air. Large particles will scatter out blue light, resulting in a pink to red sunset. Many variables contribute to a wide variety of colors, ranging from yellows to oranges, pinks, and reds.

Pink sunset light
Illuminates altocumulus globules.

Sunset yellows and oranges *over the Front Range of the Rockies. A cumulus pillar looms eastward over the mountain.*

OPTICAL EFFECTS
Sunrise/Sunset

A sunset sky *with pastel colors over the Oregon coast. Pole house pillars are in the foreground.*

Sunset colors *brighten a cirroform sky.*

A dramatic late-evening sky *over the Colorado Rockies.*

Unusual Clouds

Unusual meteorological conditions can give rise to extraordinary cloud images. They may occur at the beginning, end, or transitional point in a given cloud pattern. They can appear when two different cloud patterns merge together. Whatever might cause them, the results can be spectacular, all the more so for their rarity.

Billow

Billow clouds are organized in parallel rows. These benign-looking rows are created when wind flows at highly varying speeds in layers just above and below the clouds. These winds result in very unstable air that can be of danger to pilots. Billow clouds are most frequently seen with altocumulus and cirrocumulus.

This very dramatic sky *is packed with billow altocumulus resembling a field of rotating hawsers. The very strong shearing motion causes turbulent skies.*

Billow

Three large *"sausage rolls" of low stratocumulus soar over the treetops. Shearing forces cause rotating vortex tubes that produce clouds in the upward motion and clear space in the downward motion.*

Mostly parallel rows *of altocumulus cloud cover the sky in both these photos.*

UNUSUAL CLOUDS
Lenticular

Lenticular clouds form along mountain ranges and have very clearly defined shapes and outlines. Unlike most clouds, these stay in place, while air appears to blow through them. More accurately, cloud elements form on the uplift side, move with the wind, and evaporate on the downwind side of the wave crest. Lenticular clouds are lens shaped and form in stable air as a result of the uplift of an atmospheric wave induced by the mountain crest. As the photos here show, lenticular clouds often look like flying saucers.

A lenticular cloud
over Owens Valley,
Sierra Nevada Range,
California

UNUSUAL
CLOUDS
Lenticular

A startling *lenticular cloud appears over a mountain. Here the cloud forms in rising air over an atmospheric barrier and remains stationary.*

A fleet of lenticular altocumuli *forms over the Owens Valley, east of the Sierra Nevada Range*

High, swirling
winds create dramatic lenticular clouds, photographed at sunset at Tioga Pass in Yosemite National Park, California.

A lenticular cloud
over a mountain silhouette.

UNUSUAL CLOUDS
Cap Cloud

Cap clouds form as water vapor is lifted up on the windward side of a mountain, cools to saturation, and produces a cloud that caps the crest. These clouds are a form of lenticular cloud. Air flows through them and they remain stationary.

A cap cloud *sits on top of Mount Rainier in Washington.*

UNUSUAL CLOUDS
Cap Cloud

A cap cloud obscures *the top of Mount Everest.*

A large cap cloud *rests on the top of a snow-clad Mount Shasta, California.*

Morning sun *illuminates a cap cloud over the Andean peaks in Argentina.*

UNUSUAL CLOUDS
Noctilucent

Noctilucent clouds, as their name suggests, are clouds that seem to shine in the night sky. The word noctilucent translates roughly from the Latin as "night" and "luminous." These are extremely high clouds, forming at from 250,000 to 300,000 feet and are rarely seen. They are often vividly colored and are most likely made up of ice crystals. They form in a very dry part of the atmosphere.

An iridescent-colored *noctilucent cloud gleams in the night sky over Death Valley, California. The cloud is likely the remains of contrails created in the aftermath of a satellite launch.*

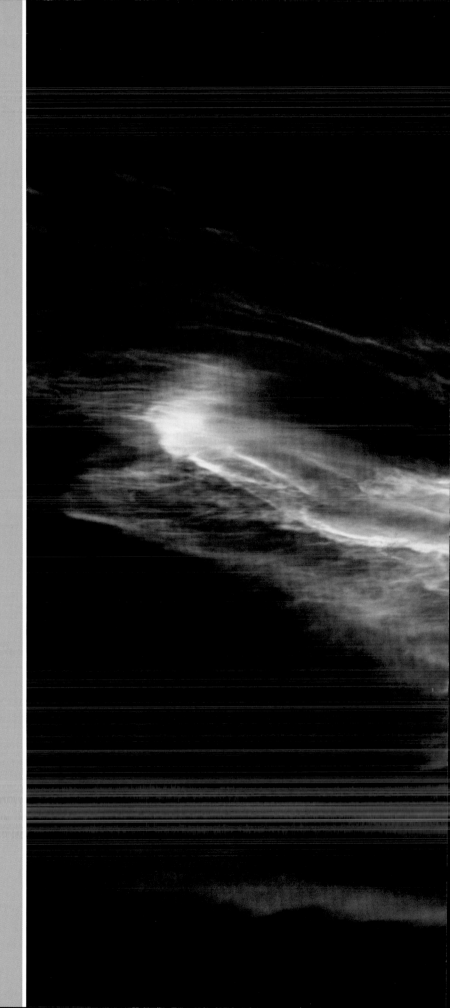

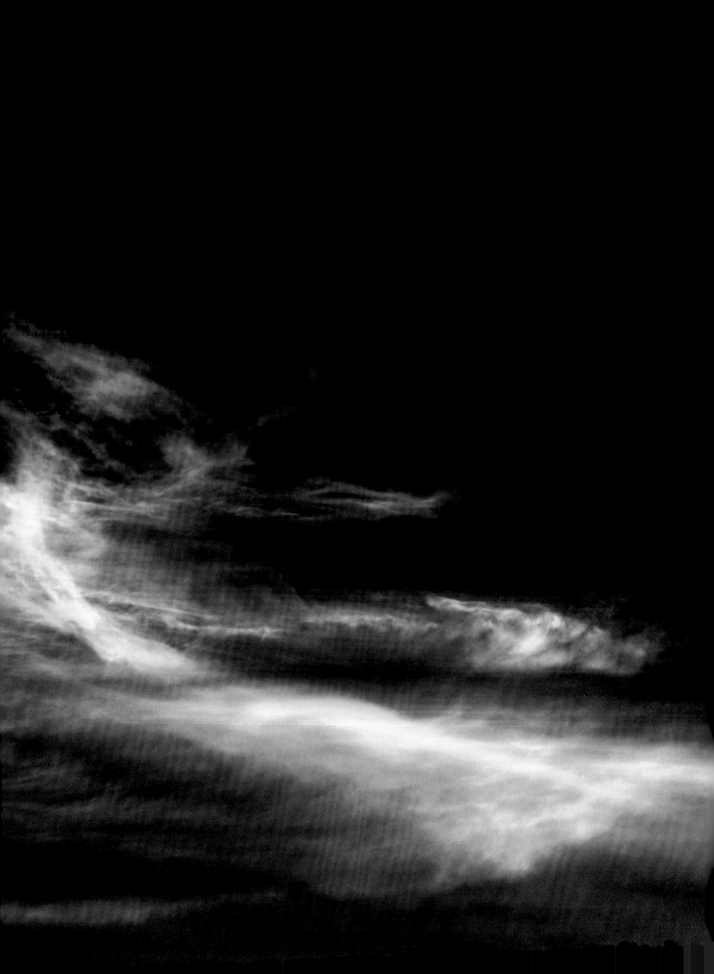

UNUSUAL CLOUDS
Nacreous

Another cloud that is rarely seen is the nacreous. This stratospheric cloud's colors have hues of mother-of-pearl and opal. The powerful iridescence can linger long after sunset. When seen at all, nacreous is visible in northern regions like Finland, Scandinavia, and Scotland. It exists at high altitudes from 70,000 to 100,000 feet and, in addition to creating a memorable display at sunset, can also be seen before dawn.

A nacreous cloud *photographed at Montrose, Angus, Scotland.*

UNUSUAL CLOUDS
Pileus

Pileus is Latin for cap. These clouds are often seen above a cumulus congestus air mass. They are formed when a layer of moist air is pushed upward over the cumulus congestus, causing condensation in this layer.

To spot a pileus *cloud, one has to look very closely. Here a delicate veil tops a rolling cumulus congestus.*

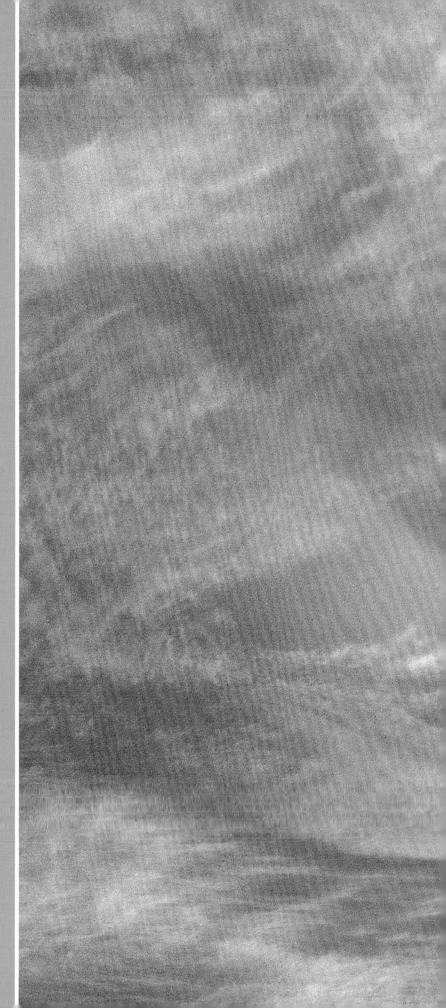

UNUSUAL CLOUDS
Hole in Cloud

A hole can be "cut" in a lower cloud, usually an altocumulus, when ice crystals from an upper cirroform cloud interact with the supercooled droplets of the altocumulus. The water droplets then evaporate and the snow crystals grow and fall, producing the hole. This is natural cloud seeding at work.

An extremely large *hole formed in a layer of billow altocumulus.*

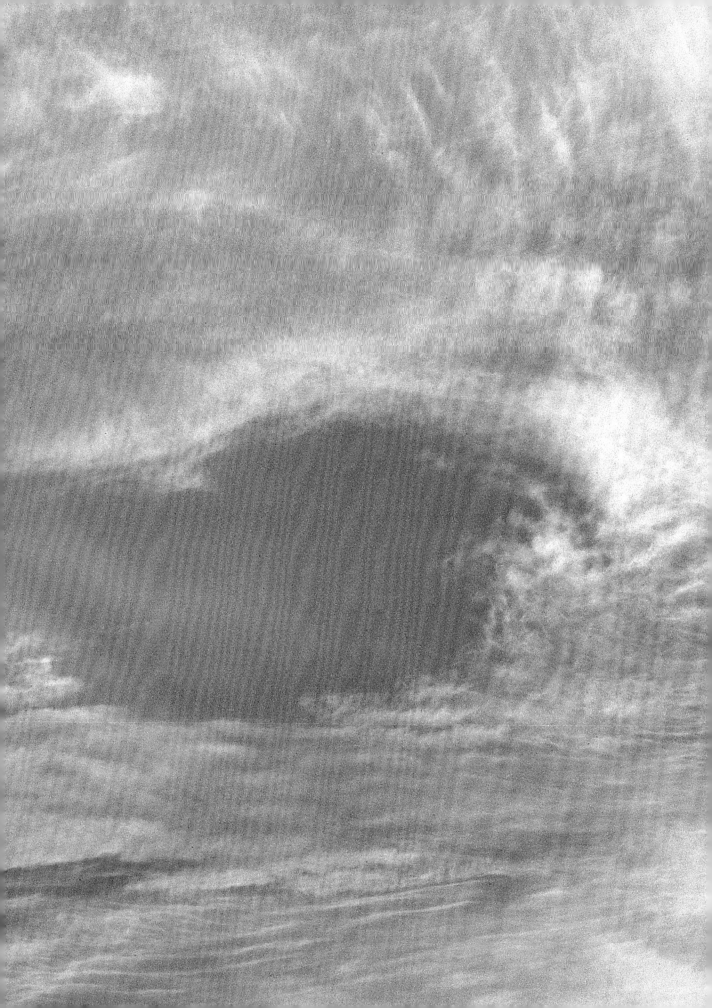

UNUSUAL CLOUDS
Sea Smoke

Sea smoke occurs when very cold air drifts across relatively warm water. This causes condensation as the warm water evaporates, and the condensation droplets form rising tendrils of fog.

Small tendrils of fog *rise from the warm water surface into the colder air above.*

UNUSUAL CLOUDS

Contrails (Condensation Trails)

Contrails are a familiar sight. They are the lines of cloud formed in the wake of high-flying jet aircraft. The jet engine's exhaust contains water vapor, which momentarily supersaturates the cold ambient air and causes condensation. Contrails can evaporate rapidly or persist and grow into sheets of cirriform clouds.

Twin contrails *have grown and formed wisps of ice crystals, called virga.*

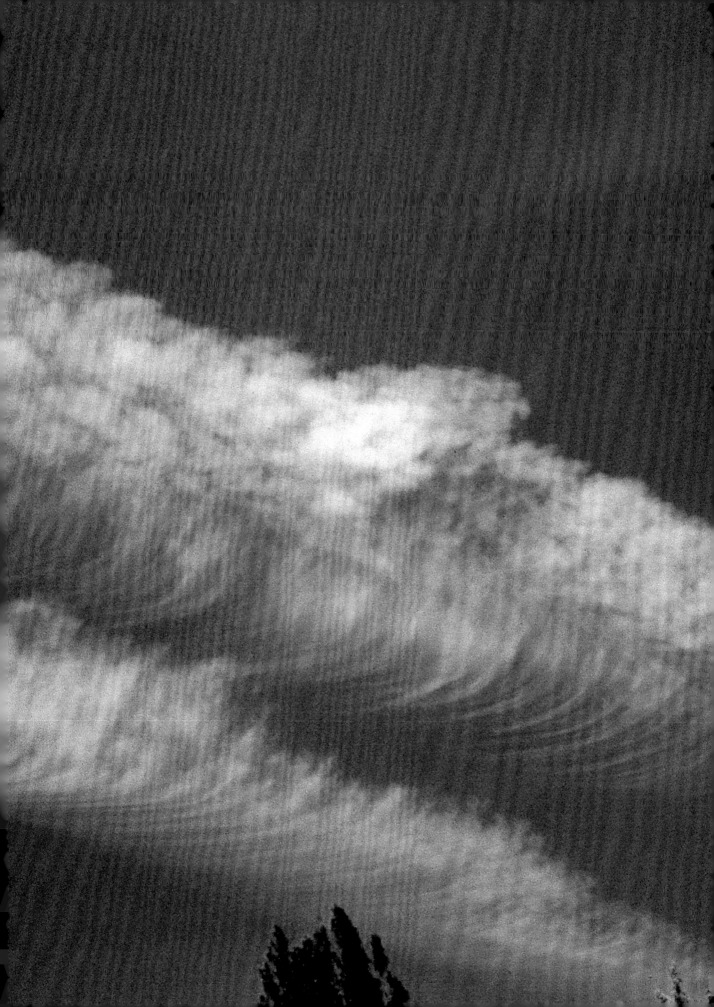

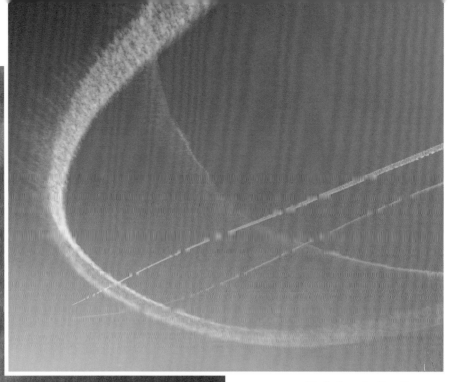

Varied contrail patterns.
The curved trail to the left shows lateral expansion and growth in a high-humidity portion of the atmosphere.

UNUSUAL CLOUDS
Contrails

Distorted tracks
of many contrails suggest shifting upper winds and an uneven distribution of water vapor.

UNUSUAL CLOUDS
Silver Lining

A silver lining cloud forms when sunlight is scattered around the edges of a dark cloud. The white rim of the cloud is caused by diffraction. Water droplets in the clouds bend the sun's light, giving a spectacular effect.

A massive, elderly *cumulonimbus is subsiding and decaying. The low sun backlights the cloud and leaves a silver lining.*

UNUSUAL CLOUDS
Funnel

A funnel cloud is associated with a tornado and can be described as the connecting link between the mother cloud and the ground. Funnels take on cloudiness when air is drawn into the low pressure vortex, then expands, cools, humidifies, and condenses.

A funnel cloud, *made visible by condensation and debris, reaches down from the mother cloud over Boulder, Colorado.*

UNUSUAL CLOUDS
Mammatus

Mammatus are globules of cloud hanging on the underside of a cloud shelf. Descending air forms the pouches. The packing of the mammatus clouds can be quite dense. These cloud forms are associated with layers of extreme instability in the atmosphere.

This array *of mammatus clouds looks like a field of tennis balls.*

UNUSUAL CLOUDS
Virga

Moisture that evaporates
as it falls from a cloud
into a layer of dry air is
called virga.

Active altocumulus
*puffballs generate rain
or snow, though it
evaporates in the
lower atmosphere.*

UNUSUAL CLOUDS
Kelvin-Helmholtz

When a warm air mass lies above colder air, the common boundary of the two is a layer of stable air. With strong wind shear, however, the interface becomes unstable, and billows of Kelvin-Helmholtz waves can occur.

A spectacular array *of Kelvin-Helmholtz waves, resembling actual waves in a surfer's paradise.*

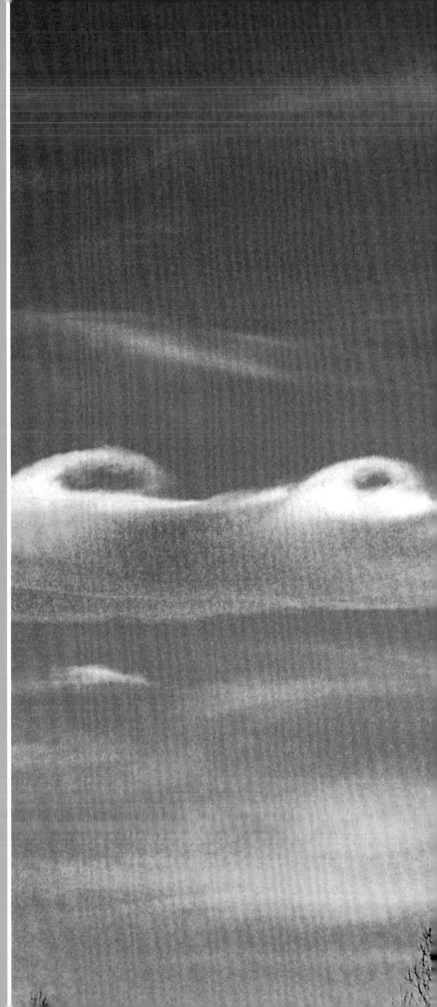

UNUSUAL CLOUDS
Interesting Shapes

The very nature of clouds promises that unusual, eye-catching formations appear with great frequency. Often these cloud forms resemble people, animals, and everyday objects. As clouds reach transition stages, new forms constantly develop, making sky watching a fascinating experience.

Although technically *this cloud formation is a decaying cumulus congestus, it is more interesting when it is viewed as a bear swimming through the air.*

LUKE HOWARD
The Man Who Named the Clouds

FOR MILLENNIA, clouds passed through the skies as interesting but nameless shapes. They were observed, painted, rhapsodized over in poems, books, plays, and sonnets. They were compared to animals and angels and described in terms of their form and color. But they had no names. Until the evening of December 1802. That night, in London, a small society called the Askesians held one of its regular meetings. Luke Howard, a young pharmacist who had been fascinated by meteorology since childhood, read a paper to his fellow members in the society. His paper was entitled "On the Modifications of Clouds." Little did any of the people in this small group know the lasting and momentous effect this paper would have on the world of meteorology. In the paper, Howard gave Latin names to the most commonly seen clouds: cumulus, stratus, cirrus, and nimbus. Those names, while amended and added to over the years, still form the basis of cloud classification today.

Crucial inventions

Why did the naming of clouds occur so late in history? After all, clouds have been influencing human behavior and activities for eons, bringing rain, blocking sun, simply being there to be seen.

The answer no doubt has to do with their ephemeral nature. Clouds appear and disappear; they change shape, move or hang still. How could such amorphous objects be analyzed and classified with any reliability? How, in fact, could scientists be expected to take these whimsical forms very seriously at all? Scientists look for order in the universe. The regular movements of heavenly bodies—stars, planets, the sun, and the moon—these were all worthy of scientific study. But a study of the capricious movements of clouds and storm systems? Not likely.

Given all that, it's understandable why astronomy

Portrait of Luke Howard, above, probably by John Opie, c. 1807.

While he did not consider himself a serious artist, Howard enjoyed painting watercolors of clouds. The images, though now faded, are nevertheless a moving tribute to his great passion.

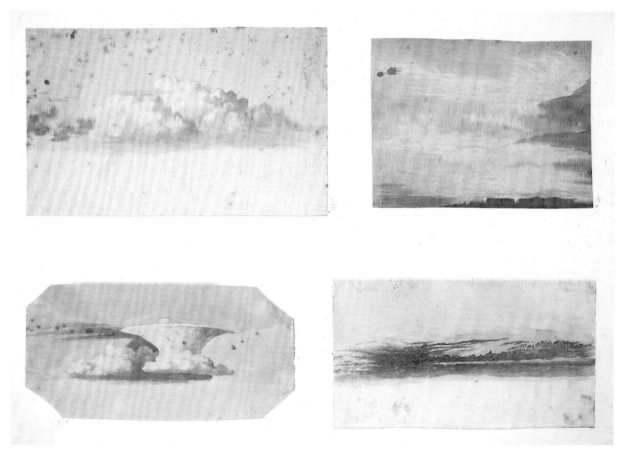

The most striking painting *here is the bottom left, a cumulonimbus mass with a dramatic anvil top on one of the cells.*

preceded the development of meteorology by centuries.

Yet in time, scientists' attention did turn to the weather around them, and to the clouds above them. As with any science, a number of inventions were essential before a systematic study could be undertaken. For meteorology, those inventions were the barometer and thermometer, which were in general use throughout the 18th century. During that century, many people with a curious nature began keeping weather diaries. By the end of the century, the laws governing the behavior of gases were becoming reasonably well understood, as was the nature of the sheath of air that surrounds the globe.

Scientists began to understand that the atmosphere did obey the laws of physics and chemistry, and that its future state could be predicted if its present and past states were accurately known. This implied careful measurement of the atmospheric variables of pressure, temperature, moisture, and wind, both at the surface and in the upper air.

But for a true ability to predict weather and weather patterns, measurements would have to be made at many places over the globe, all at the exact same moment in time. For that to happen, meteorology had to wait until the latter part of the century and the invention of the telegraph to make precise timing possible.

An early fascination

Luke Howard was born in London on November 28, 1772. His father was a staunch Quaker, and young Luke attended a Friends grammar school near Oxford. It was here that he learned Latin, more Latin there than he would ever be able to forget, he commented later. His proficiency in the subject would play a tremendous role years later when he developed his classifications.

There is no doubt that the phenomenal weather year of 1783 was an important part of Howard's early fascination with clouds. In that year the young Howard, already a devoted observer of the universe, saw nature in a state of extreme agitation and turmoil. In May and June, violent volcanic eruptions shook Iceland. The activity continued through the year, sending forth the greatest lava flow known to man. Great volumes of volcanic dust from the Eldeyjar eruption fell over Iceland and, carried by the westerly upper-level wind streams, spread over Scotland, destroying crops and killing livestock. The dust cloud eventually spread across England, continental Europe, and reached all the way to North Africa.

Meanwhile, another eruption was occurring on the other side of the world. In August, a volcano in Japan called Asamayama ejected boulders as large as houses and spewed enormous quantities of dust into the

Bands of high stratus *lie along the cliffs of a mountain range. The upper sky is filled with thick cirrostratus.*

upper atmosphere. Westerly winds carried the dust around the whole Northern Hemisphere, which added to the general pall of haze. All the weather diaries of that period and even some of the great literary works of the day contain vivid references to the extended period of the Great Fogg. The sense of underlying uneasiness was heightened by major earth tremors in Calabria and Sicily.

Finally, and most gloriously, on August 18 a large fiery meteor flashed across the skies of Western Europe. It was seen by tens of thousands of people, including young Luke Howard. There can be little doubt of the effect these events had on a ten-year-old boy who was already cloud-struck.

Observations, sketches, and a defining lecture

With his school days behind him, Howard became apprenticed to a chemist in Stockport. He endured years of drudgery as he ground chemical preparations, cleaned bottles, and swept floors. He satisfied his intellectual cravings after hours, by studying French, botany, and chemistry.

Eventually he became a pharmacist and opened his own business in London. Soon he became friends with other young men with similar interests in scientific matters. One particularly close friend was William

Allen, who owned a successful commercial pharmacy. He hired Howard to run his manufacturing laboratory in Plaistow, just outside London. As Howard traveled back and forth between London and Plaistow, he had ample time to continue his studies of cloud forms.

It was Allen who decided, in March of 1796, to establish a society for the discussion of scientific matters. This society was dubbed the Askesians, from the Greek *askesis,* which means, roughly, intellectual exercise. The meetings were held every other week at 6:00 p.m., and the rules required that each member prepare a paper on a subject of interest, and read it before the society, in rotation, or pay a fine.

While Howard was thoroughly engaged with the Askesians, he was also doing a great deal of scientific study on his own. He was particularly impressed with the work of Swedish taxonomist Carl von Linné, known more familiarly as Linnaeus. Linnaeus had established the beginnings of the modern system of classification of all life forms, a system scientists around the world would eventually adopt. Linnaeus's influence on Howard resulted in a paper Howard wrote on pollens in 1800 and read before the Linnean Society of London.

But it was the skies that continued to captivate him. Howard had presented a number of papers during the six years that he had been a member of the

Askesians, but it was the paper that he presented on an evening in December 1802 that ensured his reputation. His lecture, "On the Modifications of Clouds"—which in today's parlance would be called "On the Classification of Clouds"—was a brilliant exposition of a scheme he had developed for classifying and naming clouds. It was the culmination of Howard's years of observation. Two centuries later, we are still using the essence of his scheme.

Howard proposed in his paper that it was possible to identify a number of simple categories within the complexity of the changing skies.. He set out to establish a complete classification that would cover all possible cases. Like Linnaeus, he used Latin names, a language that fortuitously transcends national boundaries. He put clouds in four groups (see chart at left).

Cumulus (Latin for heap) Convex or conical heaps, increasing upward from a horizontal base

Stratus (Latin for layer) widely extended horizontal sheets

Cirrus (Latin for curl of hair) clouds that can stretch in any or all directions.

Nimbus (Latin for rain) Systems of clouds from which rain falls

Acclaim

One of the principal insights in Howard's essay is his insistence that clouds are a proper subject for research and theory. He noted that clouds have many shapes but only a few basic forms. These shapes and forms are caused by water present in the atmosphere, water that may be in any or all of its three states: liquid droplets, solid ice crystals, or gas (water vapor). He noted too that while circulating

A large cumulonimbus *mass diminishes the castle ruins. A heavy shower falls from the cloud base.*

air causes unstable conditions in the atmosphere, the principles of cloud formation are still understandable and are capable of being classified and studied scientifically.

Howard's other important assertion was that clouds can change from one form to another. A cumulus cloud may flatten and widen into a stratus formation. Simple observation shows that clouds move, but Howard contributed the knowledge that they change and merge. They are not fixed in one form for easy classification and study.

Even though the physics of air and water vapor were poorly understood in Howard's day, his analysis had a sound physical basis, and his paper was met with acclaim. It was soon published in Alexander Tilloch's *Philosophical Magazine,* a major dissemina-

tor of scientific knowledge. Before long Howard's paper was reprinted in a number of various journals and encyclopedias, both in England and on the Continent.

Howard continued his studies of meteorology, and his work through the first half of the 19th century included many significant achievements. In 1806, he began his Meteorological Register, which was published regularly by *Athenaeum Magazine.* He wrote the first book on urban climatology, *The Climate of London,* published in two volumes in 1818–19. In it he wrote of the impact that cities can have on meteorological conditions. For example, he carefully described London's famous fog, and noted that there could be almost no visibility in the city while a few miles away the atmosphere was clear. This seems

The cloud elements *in the cirrus watercolors seem to converge, even though they are arranged in parallel bands. In the other pictures, numerous elements of lenticular altocumulus form over the North Country hills of England.*

A larger view *of the watercolor shown at lower left, previous page. The enormous pile of cumulus congestus to the right, and the many young towers to the left suggest an unstable atmosphere and promise a thundery afternoon.*

very obvious today and not particularly noteworthy, but smog, as we call city fog today, was an unknown concept in the early 19th century.

In 1821, Howard was elected a Fellow of the prestigious Royal Society, for his contributions to meteorology.

In the second half of Howard's life, his interest shifted away from scientific matters—though his interest in clouds never flagged—and he became more involved with his chemical manufacturing business. His long life ended on March 21, 1864. At his funeral, his son said, "A beautiful sunset was a real and intense delight to him; he would stand at the window, change his position, go out of doors and watch it to the last lingering ray." Clouds in all their many moods carried an endless fascination for the man who invented their names.

Classification continues

The study of nephology (clouds) continued throughout the last half of the 19th century. Two men in particular, Professor H. Hildebrand Hildebrandsson of the University Observatory of Uppsala in Sweden and the Honorable Ralph Abercromby of the Royal

Meteorological Society, both attended the first International Meteorology conference in 1873. At the meeting there were extended discussions on an emerging comprehensive cloud nomenclature. Luke Howard had firmly established the beginnings of the classifications. Now his successors were building on his work.

Eventually, a list of ten cloud types was drawn up, expanding on Howard's four (see chart at right.)

With minor revisions, these appeared in the 1896 *International Cloud Atlas* and are in use throughout the world today.

Cumulus
Stratus
Cirrus
Nimbus
Cirrostratus
Cirrocumulus
Stratocirrus
Cumulocirrus
Stratocumulus
Cumulonimbus

Luke Howard was untrained in meteorology, a chemist and pharmacist by profession, yet through his never-ending curiosity and masterful powers of observation, he left an indelible mark on the world around him. Some call him, with good reason, the Godfather of Clouds.

Storm Clouds

LET'S GO BACK to the earlier chapter where you were spending a lazy day watching the weather in an open field. Odds are that what had started as a clear sky had changed a bit after lunch, with some clouds rolling in.

If you are fortunate, those clouds are of the cumulus variety, scattered puffs that won't amount to much. They may occasionally block the sun, causing a slight chill, but there's not enough air movement for the microscopic water droplets within them to collide or otherwise grow big enough to fall. You won't be able to wring so much as a single raindrop out of clouds like these.

But other types might give you pause. A thick gray blanket of low clouds might have scudded in, with dark, imposing undersides that promise a steady, lingering rain. Or towering clouds with anvil-shaped tops might have appeared, ominous unstable structures with lots of air movement that fosters the growth of raindrops—raindrops that will soon join their cohorts in a torrential downpour. Either way, it's time to pack up your things. A storm is coming.

Heat and Pressure

Storms are about wind and rain, of course; but before that, they are about something else—heat. Powerful amounts of energy are unleashed because of the vastly different heating that occurs between the equator and the poles. This energy results in winds, thunderstorms, hurricanes, tornadoes, and other dramatic storms across the planet.

Why the heat differential? The earth basks in the light of the sun, but it doesn't do so equally. Some parts, notably the tropical regions near the equator, receive a lot of the sun's energy, as the rays strike more or less head-on. Other regions, in particular the poles, receive a lot less because the sunlight arrives at a lower angle.

The amount of solar energy varies by season as well, because the earth's axis of rotation is inclined, rather than being perpendicular, to the sun's rays. So in July and August, for example, parts of the Northern Hemisphere—the middle latitudes from about 30 degrees to 50—are tilted toward the sun, and get far more sunlight than they do in December and January, when they are tilted away.

About half of the sunshine that enters the atmosphere is absorbed by and warms landmasses and oceans, much as sunlight hitting the skin makes it feel warm. Regions of the earth that get more direct sunlight are warmer than those that get less (that's why there are steamy conditions in the tropics and ice at the poles, after all). Warm air is less dense than colder air, because the air molecules are moving around more and fewer of them can occupy a given volume. And less density means lower pressure.

So in effect, the differences in heating around the globe create differences in air pressure. Air in the tropics tends to be warmer and of lower pressure, while air in the middle latitudes tends to be cooler and of higher pressure. Near the poles, the air is even cooler and the pressure higher still.

These pressure differences are critical. They are what cause air masses to circulate, giving life to weather.

Think of two pools of water, connected by a gate. One pool contains much more water than the other, which is thus under greater pressure. Opening the gate will cause water to flow from that pool to the one with less water. Air follows the same principle, flowing from where there is a lot of it (an area of high pressure) to where there is less (one of low pressure).

Adding a Twist

In the Northern Hemisphere, then, air tends to move between the northern latitudes and the tropics, following pressure differentials. But that's not all there is to it: the earth's spin also influences how the air circulates.

Imagine a point on the earth's surface. Like all

other points, it rotates once every 24 hours. But if that point is on the equator, in those 24 hours it has to travel much farther (about 25,000 miles, the circumference of the earth) than it does if it is midway to the North Pole (where the distance around the earth is 18,000 miles). So a point at the equator rotates at a much faster speed than one at middle latitudes, and faster still than one at the pole—which rotates in place and doesn't travel any distance at all.

When an air mass travels from the tropics northward, then, it is moving over areas that are rotating at progressively slower west to east speeds. Since the earth rotates from west to east, and the north-moving air is moving faster than the land or water beneath it, the air will curl to the east as it travels north. This is known as the Coriolis effect, and it works in the opposite direction too: air moving from the Arctic regions southward travels over areas that are moving progressively faster. So the slow south-moving air curls to the west.

It is the combination of the pressure difference and the Coriolis effect that creates a spiraling cyclonic motion. This cyclonic pattern is the major producer of storms in North America.

The Extratropical Wave Cyclone

The cyclonic effect creates what is known as an extratropical wave cyclone—"extratropical" because it occurs outside the tropics, in the middle latitudes, and "wave" because it produces a wavelike pattern

where cold and warm air masses meet.

In North America, a wave cyclone occurs when cold, dry high-pressure air, usually from the Arctic, converges with warm, moist, low-pressure air, usually from the Gulf of Mexico, along an extended line that is called the polar front. The southwestward movement of the polar air and the northeastward movement of the Gulf air create a shearing, twisting effect, deforming the front.

To the west, where the denser cold air intrudes into the warmer air, a cold front is created. To the east, where the less dense warm air is riding up over the colder air, a warm front forms. Looked at on a weather map, the two fronts form what resembles an inverted V—the wave—with zones of high pressure air on the

outside and a low-pressure zone where they converge.

This wave cyclone can sweep across the country, pushed by high-altitude west-to-east winds called the jet stream. And the pattern can change as the cold air keeps pushing into the warm air; the cold front can overrun the warm front, creating what is called an occluded front.

The wavelike deformation doesn't happen just once. It can occur over and over—the counterclockwise cyclonic motion creating a succession of wavelike cold, warm, and occluded fronts that move across the continent one after the other.

Why is this wave cyclone important? Very simply, it is along these fronts that most of our storms occur.

Frontal Storms

Along the cold front, the denser cold air pushes the lighter warm air up rapidly, where it cools and condenses into droplets, giving birth to clouds. Because of the rapid movement and the great pressure differences, the air is relatively unstable, so the clouds tend to be of the heap type—cumulonimbus storm clouds. The precipitation from cumulonimbus clouds takes the form of heavy showers or squalls, accompanied by lightning, thunder, and high-speed winds.

Along the warm front, the warm air overrides the denser cold air and rises too, though less abruptly than when forced upward by a cold front. The air is more stable, so layered clouds are the norm—nimbostratus, altostratus, or cirrostratus. In this case, the precipitation is steady.

Along an occluded front, the thick clouds and heavy precipitation of both warm- and cold-front types occur as well. Once the temperature and pressure contrasts across the fronts are reduced, the vigorous upward movement of the air matter decreases, the unsettled weather starts to subside, and clouds disintegrate. The cold, high-pressure air that sweeps in after the wave cyclone has passed to the east usually brings fairer conditions—at least until the next wave cyclone approaches.

Wave cyclones of this sort are commonplace, particularly in the fall, winter, and spring months when temperature and pressure contrasts are strongest. These contrasts produce storms that are extremely vigorous, with the strongest winds and heaviest precipitation. Storm activity diminishes in the late spring, summer, and early fall months when the differentials are much less dramatic.

Convective Storms

The arrival of a cold or warm front is not the only reason that air ascends. Air will rise when it is heated—when cold air passes over a land area that has been warmed by the sun, for instance. The heat from the ground is transferred to the air, which decreases the density of the air. This less dense air rises and is replaced by cooler, more dense air. The process is called convection.

Weak convective updrafts often result in fluffy fair-weather cumulus clouds; but with strengthening updrafts, these innocuous clouds can develop into the more threatening cumulonimbus variety, bringing a rainstorm.

Thunderstorms

A look at the interior of an active cumulonimbus cloud shows a strong updraft in the core, with secondary downdrafts and a generally chaotic array of up-and-down motion of cloud droplets and growing raindrops. The top of the cloud can be high enough to have freezing precipitation such as ice crystals, snowflakes, and frozen rain.

There is also electrical activity within. To understand this, consider first that the atmosphere contains electrically charged gases called ions. With all the chaotic air movement, cumulonimbus clouds can become polarized, like a battery. There is a concentration of positive electrical charges at the top of the cloud and negative charges near the base.

The earth's surface has electrical charges too—normally, negative ones. Since like charges repel, the effect of the negative charges at the bottom of the cloud is to repel the negative charges on the ground. This leaves the ground near the cloud positively charged.

The contrast between the negatively charged cloud base and the positively charged ground creates an electrical potential between the two. If the charges build up, at some point current will flow. The air, which normally acts as an insulator, suddenly becomes a conductor of electricity. A surge of electricity passes upward and downward between the cloud and the ground. This is the flash we call lightning.

The temperature in the immediate vicinity of a lightning bolt is about 50,000° F. This causes abrupt heating in the adjacent air, sudden changes in atmospheric pressure, and finally a pressure wave that travels outward at the speed of sound, causing a thunderclap. Sound takes about five seconds to travel one mile, but light reaches us almost instantaneously. So if, for instance, we hear thunder fifteen seconds after seeing lightning, the flash occurred three miles away. When the lightning is very close, we hear a loud clap of noise. When it is far away, thunder seems to rumble.

Hailstorms

Hailstorms are a warm-weather phenomenon, caused by a severe cumulonimbus cloud that reaches high into the atmosphere, creating cold, turbulent conditions within.

The hailstones themselves are ice particles that are transparent, or mostly so, and range in size from seeds to baseballs. The largest authenticated stone in the United States, one that fell in Nebraska in 1928, was almost six inches in diameter.

Many hailstones form in layers. Opaque layers are formed when the stone encounters a region of many small supercooled droplets that freeze on impact with it and trap air between them. Clear layers are a result of the stone's encounter with larger supercooled drops that freeze more slowly. The turbulent updrafts in a severe cumulonimbus can keep a hailstone growing until it is very large and the updrafts can sustain it no longer.

Tornadoes

The cumulonimbus is an awesome cloud, which grows in very unstable air marked by a strong updraft. A single cumulonimbus, called a cell, is rare however; normally the turbulent mixing of warm and cold air will help give rise to other cells nearby. What we consider a storm cloud is usually a cluster of cells in various stages of development.

If a cluster grows very large, it's called a supercell—a very dangerous storm with a violent updraft and, quite possibly, a vortex of spinning air within that begins to take on a life of its own. A supercell can be, literally, a factory of extreme weather. It's the spawning ground of tornadoes.

The tornado is without doubt the most awe-inspiring (and fear-inducing) atmospheric phenomenon found in the middle latitudes. Its distinguishing feature is a funnel cloud, which extends earthward from the base of the cumulonimbus cloud. The funnel can vary in shape; it can be a narrow inverted cone, a

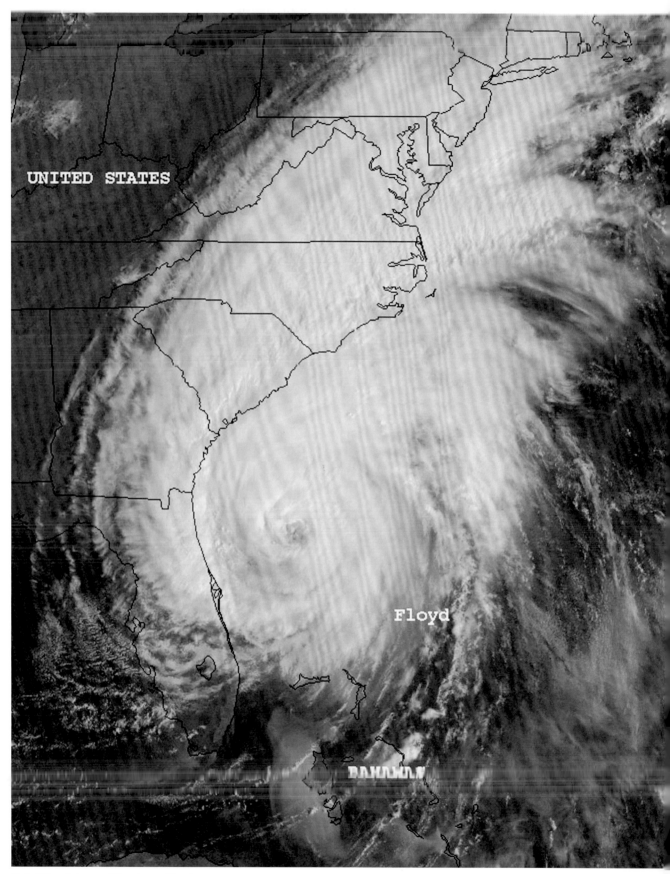

Satellite picture *of Hurricane Floyd*

vertical cylinder, an hourglass, or a long, twisting rope. Sometimes the funnel is distinct, and at other times it is obscured by dust and rain squalls.

As the central pressure in the funnel drops, air that is sucked into the vortex cools by expansion and reaches saturation. Then condensation forms clouds that outline the funnel shape of the vortex. As droplets grow larger and dust and debris are drawn into the vortex, the funnel takes on its dark and ominous appearance.

A funnel cloud can move at an average speed of 30 to 40 miles per hour—fast, but not fast enough to do much damage. It is the speed of the winds within and adjacent to the funnel that can cause horrific destruction. Because of the excessively strong differences in pressure around the low-pressure core, these winds are unbelievably strong, reaching 300 to 500 miles per hour or more.

The width of a funnel cloud rarely exceeds one mile and is more often only a few hundred feet. The length of a tornado's path before it dissipates is usually not much more than 10 miles and can be much shorter. That's why a tornado can hit one town hard and spare others nearby. Even in a neighborhood that's been hit, a tornado seems wickedly fickle. The houses on one side of a street might be destroyed while those on the other side are unscathed.

Tropical Circular Storms

The extratropical wave cyclone is North America's major weather maker, but it's not the only one. In the late summer and fall, storm systems that form in the tropics can reach the continent and wreak havoc as well. At their worst, such storm systems are called hurricanes in the Atlantic area, typhoons in the northern Pacific, and willy-willies in the South Pacific.

A young storm of this type is called a tropical depression, and it usually forms in the tropical Atlantic or Pacific. As this low pressure wave, or trough, slowly progresses toward the west, driven by trade winds, the disturbance can intensify and develop a cyclonic pattern of winds. If conditions are right, it can become a tropical storm—defined as a cyclonic storm with winds of up to 73 miles per hour. If it continues to grow in strength as it reaches the western end of the Atlantic basin, so that winds exceed that speed, it is considered a full-blown hurricane.

When a tropical system, whether a hurricane or some lesser storm, reaches the western end of the

Atlantic, the track usually shifts and the storm recurves, bringing with it warm, fast-moving tropical air that contains tremendous amounts of moisture. These tracks run the range from predictable to completely unpredictable and erratic. Some feint north, double back to the south, move westward into the Caribbean, and make landfall in Mexico or along the tier of southern states. Others track inland along the heavily populated eastern seaboard.

Hurricanes are the unstoppable freight trains of weather, and their freight is wind and water. In just a day or two, a hurricane can dump many inches of water on a region, causing extensive flooding. The high winds can uproot trees and rip roofs off buildings. Storm surges—high, violent waves and tides that accompany a hurricane—can destroy beach communities.

Since hurricanes are kept under 24-hour surveillance from weather satellites and aircraft, weather forecasters can give quite accurate information about expected tracks and changes in intensity. But hurricanes have been known to shift course unexpectedly, and that's why evacuations are occasionally ordered for wide areas of the East or Gulf coasts. Forecasters may not always know exactly where a hurricane will hit land, but they do know that wherever it hits, the damage will be extreme.

*   *   *

Hurricanes and tornadoes lie at one end of the storm spectrum; fluffy little cumulus clouds that will never amount to much are at the other. In between, there are weather possibilities galore, including ice storms, windstorms, and dust storms, to name just a few.

Storms start on their own terms, and stop on them as well. Some end when enough air movement has caused the pressure and temperature differences that created them to finally subside. Some are broken up by winds. But others simply run out of moisture or burn up all their wind through friction with the ground.

Unfortunately for us, a storm that has dumped all its rain or used up all its wind has usually left a mess in its wake. The extent of the damage depends in part on whether we heeded forecasters' warnings and took precautionary measures, or ignored them and suffered the consequences.

Forecasting Weather

ANYONE CAN PREDICT THE WEATHER. You've done it yourself, no doubt—that time you left the house, saw a few thunderheads in the distance, and turned back to fetch an umbrella, for instance. Or that winter morning when you saw a halo around the sun and promptly went to the hardware store to finally replace that battered old snow shovel.

Sure, as a modern society we've lost much of our connection to the earth. We ride around in cars, eat food that has been produced and packaged for us, work in air-conditioned offices, and sleep in air-conditioned houses. We're more likely to watch the Weather Channel than to spend an afternoon out in the elements watching the weather.

But to some degree, most of us still have an innate feel for our immediate environment. We can recognize that when the wind freshens up, a front is moving through; or that a layer of morning fog is going to burn off to make a beautiful day.

In this, we are continuing a tradition that began with the very first humans. As long as people have walked the earth, they have been reading the skies. Often their livelihood, if not their lives, depended on their ability to predict coming conditions based on what they saw. They looked for signs, and most of those signs were clouds.

Among the earliest documented examples of forecasting were the efforts of the Babylonians, who in the 7th century B.C. used clouds to foretell weather changes over the short term. Several centuries later, Aristotle wrote treatises on the weather (although he got much of it wrong).

For centuries, weather forecasting was almost all art and little or no science. Often it was little more than folklore, like the adages about the weather at sea passed down by generations of mariners (and occasionally still heard today). One example that uses clouds as an indicator:

> When clouds are getting thick and fast,
> keep sharp lookout for sail and mast
> But if they slowly onward crawl,
> out with the dories, nets, and trawl.

Or this one:

> Horses' manes and mares' tails,
> Sailors soon shall shorten sails.

Forecasting Science

In modern times there's still some art that goes into making a useful forecast. But predicting the weather has become much more scientific.

Science needs tools for gathering information, and for the science of weather forecasting. Among the first tools were instruments like the thermometer, hygrometer, and barometer, invented in Europe during the 15th to the 17th centuries and used to measure temperature, humidity, and air pressure.

But even with these tools, those who wanted to understand the weather were hindered by a simple fact: they couldn't be in more than one place at a time. An observer could measure conditions in one location, but since the only means of communicating was through travel, there was no way of knowing what conditions were like elsewhere at the same time.

That changed in the mid-19th century with the invention of the telegraph, which might be considered the single most important development in the history of weather forecasting. Now observers could learn about conditions in many places at once. By recording and plotting the information on maps, they could begin to see patterns in the weather.

Weather science, or meteorology, blossomed in the late 19th century and continued throughout the 20th. The nature of the atmo-

Evangelista Torricelli *(1608-1647),*
the inventor of the barometer.

sphere and the complex processes that take place within it became a rich field for research. Scientists became aware that the atmosphere, the land, and the sea were inextricably linked; that each affected the other, and that the sun affected all three.

They knew that in order to forecast weather, they had to understand the variables of temperature, humidity, wind, and pressure. To do that, they had to know what was going on with the weather, at any given moment, around the globe. For meteorologists, then as now, no amount of information was too much.

Networks of ground-weather stations were established. Instruments were sent aloft by balloon and airplane. Later, a system of ocean buoys was developed and deployed to gather data. Now, in the space age, weather satellites provide tremendous amounts of real-time information about cloud forms and weather.

Along with vast increases in information have come great advances in how it is analyzed. We understand much more about the way weather works, in large part because we can now model it mathematically, expressing the physical laws that govern the workings of the atmosphere as equations and plugging weather data into them.

In this, a pioneer was Lewis Richardson, a British mathematician. His 1922 book, *Weather Prediction by Numerical Process,* explained how the use of grids of weather data entered into mathematical equations could produce expected values for variables like temperature and pressure at some future time. In other words, mathematical analysis could predict the weather.

This involved an immense number of calculations, and Richardson had the misfortune of living in the age before high-speed computers. Working by himself, he took several months to come up with a six-hour forecast (one that, as it turned out, was not very accurate). He then developed a method by which human "calculators," each working on part of an equation, could do the work faster. The only problem was that he'd need 64,000 people, all in close proximity so that they could pass their results on to the next person in the process. Obviously that was impossible.

It was only with the development of computers in the second half of the 20th century that mathematical weather analysis became feasible. By 1950, the first successful 24-hour forecasts were made by computer, and as machines have become more powerful,

forecasts have become more accurate and finely tuned. There is so much weather data, and so many ways for it to interact, that some of the most powerful computers ever built—so-called supercomputers—have been called on to help.

Forecasting Art

Still, forecasts can be notoriously inaccurate. The major problem is that they all extrapolate the information into the future, and even the best models and fastest computers cannot take into account every conceivable possibility. Often human judgment, based on knowledge and experience, can be as useful as the calculating ability of a machine.

If a forecaster has access to weather data that has been analyzed and plotted on a map, the positions of low- and high-pressure zones and associated cold and warm fronts over the previous several days can be studied. Then, in the simplest case, a forecaster projects ahead, extrapolating from the data to calculate the future positions of the fronts and pressure zones, adjusting them based on a judgment of how much the weather system will accelerate or decelerate. Knowing cloud and precipitation regions and associated winds, a forecaster projects the future position of these as well and can then come up with a forecast for a particular region.

The problem with this procedure comes when dealing with change. Is the low-pressure center deepening or filling? Is it speeding up or slowing down? A forecaster's skill at evaluating these factors determines the accuracy of the forecast. Many times weather systems that seem well behaved will suddenly and unexpectedly stall or speed up, turning what had seemed like a perfectly respectable forecast into a bust.

Today, local television and radio stations often give a 10-day weather outlook, but these contain more bluster than accuracy. The state of the art of forecasting, as sophisticated as it has become in recent years, simply does not possess that capability. The American Meteorological Society, for example, says that accuracy at forecasting temperature decreases from "good" at three days to "fair" at five days; for precipitation, accuracy is "fair" at three days and "marginal" at five days. Still, all the technology and experience has led to improvements: five-day forecasts today are considered roughly as accurate as three-day forecasts were a decade ago.

Aerographers release a weather balloon.

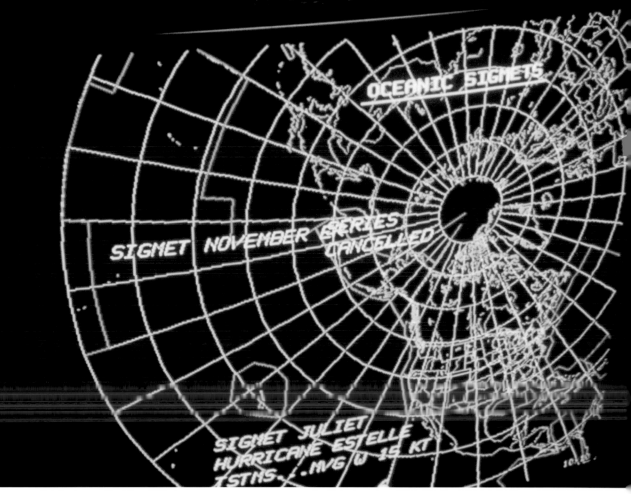

A colored weather map *at the Federal Aviation Administration Air Traffic Control Center in Fremont, California.*

Forecasting for Yourself

Most people don't have either the technology or the experience to make detailed and accurate predictions of the weather. But with a little bit of help from very basic technology—a barometer—and a little bit of knowledge about clouds, you can make some good guesses, particularly about what the next day or so has in store.

Reliable Weather Indicators

1. Conditions:
   Cirrostratus clouds with halo, thickening to altostratus; southeast wind increasing in speed and veering to the south; barometer falling.
   Forecast:
   Continuous rain in 12 hours.

2. Conditions:
   Cirrostratus with halo; southeasterly winds backing to east. Barometer steady or rising.
   Forecast:
   Becoming more cloudy with light rain in 24 hours.

3. Conditions:
   Partly broken to broken stratocumulus; continuous rain ceasing; southwesterly winds moderate; warm and muggy.
   Forecast:
   Cold front passage in 12 hours, with heavy showers.

4. Conditions:
   Heavy showers with wind shifting from southwest to northwest; barometer rising.
   Forecast:
   Clearing weather with cooler temperatures.

5. Conditions:
   Cumulus congestus clouds, thickening with towers rising.
   Forecast:
   Showers within hours.

6. Conditions:
   Altocumulus clouds forming by midmorning; hot and humid.
   Forecast:
   Thunderstorms by afternoon.

7. Conditions:
   High cirrus clouds in long tails, or altocumulus in a pebbled pattern.
   Forecast:
   Warm front approaching; deteriorating weather.

8. Conditions:
   Light cumulus, well scattered.
   Forecast:
   Fair weather continues.

9. Conditions:
   Cloud cover at night, cold temperatures.
   Forecast:
   Warming tomorrow.

10. Conditions:
    Lightbulb-shaped darkening and lowering cumulonimbus, barometer falling rapidly.
    Forecast:
    Severe storm with high winds possible soon. Seek shelter.

Watching the clouds is a far cry from crunching satellite data in a computer. But it's worked for centuries, so there's no reason why it shouldn't still work today. A cloud may not be the most accurate predictor of weather, but more often than not it can provide useful clues. And if you are away from a radio or TV—remember that picnic in the field?—it may be the only way to figure out that the rest of the day is going to be clear and sunny. But here's another useful rule: it never hurts to pack an umbrella, just in case.

Observing Clouds

Looking up with an analytical eye

For perfect weather observation, a scientist would have to gather complete information about the state of the weather in every spot on the globe. This would include observations about the cloudiness in the sky, air pressure, temperature, and humidity. Obviously, with so many areas of the earth unpopulated, an unbroken network of weather stations is not feasible.

The desire to secure complete data on the weather has gone on for a long time, as probing the depth dimension of the atmosphere has always presented a great logistical and technological challenge. Finally in the first half of the last century, after a series of balloon and airplane probes of the upper atmosphere, a system called rawindsonde was introduced. Rawindsonde is a system that consists of large helium/hydrogen–filled balloons that carry instruments for measuring wind, temperature, and humidity. These measurements are then radioed to ground stations. Rawindsonde was a great leap forward, since it provided more information than we had before, but scientists wanted still more in-depth data.

Great strides were made in April 1960, when a polar-orbiting satellite called TIROS-1 was launched into space. An acronym for Television and Infra Red Operational Satellite, TIROS-1 was fitted with visual and infrared cameras and sent back pictures of the earth's cloud cover that thrilled the meteorological community. These soon became a principal aid of weather forecasters.

In the 40 years following TIROS-1, there has been an explosion of satellite technology. In addition to relaying visual images of cloud cover, satellites provide platforms for sophisticated instruments. Some of these instruments are atmospheric sounders that measure clouds and water vapor at various levels in the atmosphere. Over the period of time it takes for the earth to make one revolution, data gathers to fill every spot on the entire planet. These measurements fill in the large gaps that exist between weather stations in the rawindsonde network.

A weather satellite (*left*) *mounted on a rocket for launching into earth orbit. Above, a satellite view of an Atlantic hurricane approaching the southeast coast of the U.S.*

On May 4, 2002, as a part of NASA's Earth Observing System, another important tool was created when the Aqua spacecraft was launched from Vandenberg Air Force Base in California. Among the instruments aboard Aqua were two CERES (Clouds and Earth's Radiant Energy Sensors). A principal objective of the launch was to provide the NASA CERES science team with data to investigate the role of clouds and water vapor in the heating and cooling of the planet. We know how important the role of clouds is in determining our climate. For example, a cloud cover can warm the earth by trapping heat beneath it; under other circumstances, it can cool the earth as it reflects sunlight from the earth back to space. But we need more data to understand clouds' total import. For instance, how much heat do clouds actually absorb? How exactly do they react to changes in temperature? The telescopes that make up CERES should answer some of these questions.

NASA has enlisted the help of schoolchildren to find some of these answers through a program called S'COOL (Students' Cloud Observations On-Line.) Schools that become part of this program have their students make ground measurements for the CERES experiment. As the satellite passes over them, the students make their basic weather observations and record the type and features of clouds they see in the sky. Their measurements will be compared with the data the Aqua satellite provides from above.

The observation of clouds, then, can range from a quintessential pleasant pastime to a vital government experiment.

Photographing Clouds

Photography involves seeing through the lens of a camera. But truly seeing has many different dimensions. We see with the physical eye, but also with the inner eye—the eye of the artist that is innate in each individual but untouched in most. When you see with the inner eye, each cloudscape is unique and worth savoring. Some cloudscapes feature such an array of color and arrangement of form that the result is beauty almost beyond description.

The following guidelines should prove useful in photographing clouds.

1. Keep your camera rock-solid when you press the shutter release. If possible, use a tripod, or rest the camera against a solid object. Practice holding the camera firmly and moving only the finger that activates the shutter release. Use of fast ASA film will minimize this problem.

2. Use a haze or sky filter continuously.

3. Use a polarizing filter continuously. Using this filter increases the contrast between clouds, particularly those in the cumulus family, and the background sky, thus enhancing the cloud image. Polarized light maximizes at 90 degrees to the solar beam, as you will find by pointing the camera to various parts of the sky. While most point-and-shoot cameras will not accommodate screw-on filters, a filter can be held in front of the lens and rotated to produce the desired effect. Take care not to allow a stray finger to impinge on the incoming light.

4. Use a neutral density filter when a bright sky and dark foreground are juxtaposed.

5. Become aware of the subtleties of light. In general, avoid photographing in the harsh light of midday. There is more drama in the light of the low sun of morning and late afternoon.

6. Learn the art of composition. The artistic value of a cloudscape often is determined by the arrangement of the various elements that comprise the final image. As a first step, eliminate intrusive foreground material such as bushes, trees, light wires, and poles. Tree branches sometimes come in handy as frames for a photograph.

7. Most cameras used by amateurs include an automatic-focus capability that is activated by an infrared beam. If you have the ability to set the f-stop, you can decide on the depth of focus you want, in order to enhance or obscure particular elements in the scene you are shooting. Fortunately this is only a minor consideration when shooting cloud images because there are usually no very close-up objects in the picture.

8. When photographing a halo or corona around the sun, find some object with which to block out the solar disk. Never look through the viewfinder directly at the sun. That could cause eye damage.

9. Always remember that clouds are ephemeral, always changing, disappearing into invisible vapor and reappearing in visible form. This is nature's "now you see it; now you don't" magic act. So click and capture the moment.

10. There are several brands of excellent quality film now on the market, led by Kodak, Fuji, and Agfa. As for speed, ASA 200 is a good choice for most photographers of the sky.

11. Hone your skills by practicing, and by stopping to think before you click. Is your inner eye awake?

12. Study cloudscapes taken by master photographers like Ansel Adams. Try to think as an Adams. Who knows, you may have the potential to become another master photographer of the skies.

Glossary

Adiabatic process A process of heating and cooling that occurs without an exchange of heat between the system (such as an air mass) and its surroundings. In an adiabatic process, warming comes from compression; cooling comes from expansion.

Advection The horizontal transfer of any atmospheric property by the wind.

Advection fog Fog that occurs when warm, moist air moves over a cold surface and the air cools to below its dew point.

Aerosols Tiny solid particles (of dust, smoke, etc.) or liquid droplets that enter the atmosphere from either natural or man-made sources.

Air mass A large body of air that has similar horizontal temperature and moisture characteristics.

Altocumulus A mid-level layered cloud that indicates instability in the atmosphere. Altocumulus clouds are sometimes arranged in rows or small ripples.

Altostratus A mid-level cloud that produces overcast weather conditions. It can sometimes be so translucent that the sun shines through it. Altostratus clouds generally mean that rain is imminent.

Altostratocumulus A mixture of layer and heap clouds that form in the middle atmosphere.

Anticrepuscular rays These rays appear on the antisolar point, which is the point on the sky that is opposite the sun. These high parallel shafts of light, through a trick of perspective, seem to converge in the distance. (See also "crepuscular rays")

Atmosphere The gases, mainly nitrogen and oxygen, that surround the earth and are held to it by gravity.

Aurora borealis Curtains of light that occur in the high-latitude upper atmosphere of the Northern Hemisphere. The flashes are caused by charged particles from the sun that interact with the earth's magnetic field. The magnetic field disperses the particles to the polar regions, where they react with gases in the earth's upper atmosphere to cause the glowing light display.

Barometer An instrument that measures atmospheric pressure.

Billow clouds Broad clouds organized in parallel rows. The rows are created when wind flows at varying speeds in layers just above and below the clouds.

Cap cloud This type can form when water vapor is lifted up on the windward side of a mountain, producing a cloud that hangs like a cap on the mountaintop. It is a form of lenticular cloud.

Cirrocumulus A high-level cloud that looks like an array of small balls, often organized in rows. Their appearance brings the promise of storms.

Cirrostratus Very high, thin sheetlike clouds composed of ice crystals. They frequently cover the entire sky and often produce a halo around the sun or moon.

Cirrus A high-level cloud with a varied wispy appearance. It is generally white with a silky sheen and is composed of ice streamers. The ice crystals usually evaporate before reaching the ground.

Coalescence The merging of cloud droplets into a single larger droplet.

Cold front A boundary line along which a cold air mass advances and pushes a warm air mass rapidly upwards.

Condensation The process in which water vapor becomes a liquid. In weather, this would be the point at which the atmosphere is filled to overflowing with moisture.

Condensation nuclei Tiny water-loving particles composed of dust or salt or aerosols, with slight charges that attract and hold on to water particles.

Contrail A ribbonlike cloud that frequently forms behind jet airplanes flying in clear, cold, humid air.

Convection Motion in the atmosphere that occurs when heat from the ground warms the air, which rises and is replaced by cooler, denser air. Convective updrafts result in cumulus-type clouds.

Convergence An atmospheric condition that occurs when several airstreams moving horizontally arrive at the same place. Some of the air must rise, since there is not enough space for all of it in the region where the airstreams meet.

Coriolis force An apparent force that results from the rotating of the planet. Moving objects, such as the wind, are deflected to the right of its motion in the Northern Hemisphere and to the left in the Southern Hemisphere.

Corona A series of colored rings that appear to be centered around and close to the sun or moon. They are caused by the diffraction of light as it passes through a cloud of small droplets or ice crystals.

Crepuscular rays Alternating dark and light bands of light that seem to radiate fanlike from the sun. (See also "anticrepuscular rays")

Cumulonimbus A very dense, towering cloud whose top is often in the shape of an anvil. Heavy showers often fall from the base, as can hailstones. This cloud is also associated with displays of lightning and may produce tornado funnels.

Cumulus A cloud with a flat bottom and a rounded top that resembles cauliflower. Cumulus clouds appear in a stable atmosphere and are usually associated with fair weather, although unstable air can lead to cumulus congestus clouds.

Cumulus congestus The last stage in development of the cumulus family prior to the cumulonimbus stage. The cloud's multitude of cells is in all stages of development due to convective heating. Light showers can fall from these clouds.

Cyclone An area of low pressure around which the winds blow counterclockwise in the Northern Hemisphere and clockwise in the Southern Hemisphere.

Dew point The temperature to which air must be cooled for saturation to occur.

Diffraction The bending of light around objects such as clouds and fog droplets, producing fringes of light and dark or colored bands.

Drizzle Small water drops between 0.2 and 0.5 millimeters in diameter that may appear to float and that reduce that visibility more than the larger drops of light rain.

Evaporation The process by which a liquid changes into a gas, or vapor.

Fair-weather cumulus Widely spaced low-level puffs of cloud that resemble sheep. They usually form on summer afternoons.

Fog A cloud with its base at or near the earth's surface.

Front The boundary between two distinct air masses.

Funnel cloud A rotating, conelike cloud that extends downward from the base of a thunderstorm. When it reaches the ground it is called a tornado.

Glory A complex optical phenomenon that is seen when an object's shadow is projected on a cloud.

Green flash A momentary flash of brilliant emerald-green light that appears just above the sun as it drops below the horizon. It is best seen over the sea in a stable atmosphere.

Ground fog Fog produced over the surface of the earth when the temperature drops to the dew point and the air becomes saturated. Further cooling changes the water vapor into tiny droplets that comprise the fog.

Hailstones Transparent or partially opaque balls of ice that range in size from a pea to, on occasion, a baseball or even larger.

Halo An optical effect of a ring or arc that results from the refraction of sun- or moonlight passing through tiny ice crystals in the atmosphere.

Hole in cloud A hole cut in a lower cloud, usually an altocumulus, when ice crystals from an upper cirroform cloud intersect with the supercooled droplets of the altocumulus. The water droplets then evaporate and the snow crystals grow and fall. The process is similar to clouds that are deliberately seeded to produce rain.

Humidity The water-vapor content of the air. (See also "relative humidity")

Hurricane A severe tropical cyclone with winds in excess of 74 miles per hour.

Hydrologic cycle The movement and exchange of water among the earth, the atmosphere, and the oceans as it evaporates, condenses to forms cloud, and falls as precipitation.

Hygrometer An instrument that measures the water-vapor content of air.

Inversion An increase in air temperature as air rises, rather than the more usual situation, which is an increase when air falls.

Iridescence Pastel colors, sometimes similar to the hues found in mother-of-pearl, that are seen in clouds when drops of water that are the same size diffract sunlight. The colors usually appear at the edges of thin altocumulus clouds.

Jet stream Strong winds concentrated within a narrow band in the high atmosphere. The winds blow generally from west to east.

Kelvin-Helmholtz This cloud formation can occur when a warm air mass lies above colder air, with a common boundary of stable air. If strong wind shear makes that interface unstable, billows of Kelvin-Helmholtz waves may appear.

Latent heat The heat that is either released or absorbed by a substance when it undergoes a change of state, such as during condensation (when water vapor becomes liquid) or evaporation (when liquid becomes a gas, or vapor).

Lenticular cloud A lens-shaped cloud that forms in stable air. These clouds often look like flying saucers.

Lightning A visible electrical discharge produced by thunderstorms.

Low stratus These layered clouds are formed when weak upward air currents lift a thin layer of air high enough to begin condensation. Light mist can fall from these clouds.

Mammatus clouds Globules of dense cloud hanging on the underside of a cloud shelf. These

cloud forms are associated with layers of extremely unstable air.

Molecule A collection of different kinds of atoms held together by electrical forces.

Nacreous clouds Infrequently seen clouds that have hues of mother-of-pearl and opal. They form at very high altitudes, and their powerful iridescence can linger long after sunset.

Nimbostratus A mid-level cloud from which continuous rain or snow falls.

Noctilucent clouds Extremely high clouds, forming at from 250,000 to 300,000 feet. Wavy and thin, they are often vividly colored and are most likely made up of ice crystals.

Occluded front A complex frontal system that forms when a cold front overtakes a warm front.

Orographic lift The lifting of air over a physical barrier, such as hills or mountains. The lift is the main reason clouds and precipitation tend to form on the windward sides of mountains, such as the slopes of the western Rockies.

Parhelia Optical effects seen in many forms, including circles around the sun or moon or bright spots on either side of the sun or moon. They result from the refraction of light through tiny ice crystals in the atmosphere. They include sun dogs and sun pillars.

Pileus A smooth cloud in the form of a cap. These clouds form when a layer of moist air is pushed upward over a cumulus congestus cloud.

Polar front A semipermanent, semicontinuous front that separates tropical air masses from polar air masses.

Precipitation Any form of water particle, liquid or solid, that falls from the atmosphere and reaches the ground.

Rain Precipitation in the form of liquid water drops that are greater than 0.5 millimeters.

Rainbow An arc of concentric colored bands that is formed when light from either the sun or the moon is refracted and internally reflected in shower drops.

Rawinsonde A balloon-borne system that carries instruments aloft for measuring wind, temperature, and humidity.

Reflection The process in which a surface turns back a portion of the radiation that strikes it.

Refraction The bending of light as it passes from one medium to another of different density.

Relative humidity The ratio of actual water vapor present as compared to the maximum amount possible at a given temperature.

Saturation A condition in the atmosphere in which the maximum amount of water is present, relevant to a given temperature and pressure.

Sea smoke This cloud forms when very cold air drifts across relatively warm water. This causes condensation as the warm water evaporates, and the condensation forms into rising tendrils of fog.

Shower Intermittent precipitation from a cumulonimbus cloud, usually of short, heavy duration.

Silver lining A bright border that forms when sunlight is scattered around the edges of a dark cloud.

Sleet Also called ice pellets. Precipitation that consists of ice pellets 5 millimeters or less in diameter.

Snow Frozen precipitation composed of ice crystals in complex hexagonal form.

Stable air A condition in the atmosphere that exists when lifted air is colder than the air around it.

Stratocumulus A common low-level layer cloud, with lumpy, rounded elements.

Stratosphere The layer of the atmosphere above the troposphere, generally beginning about 7 miles up and extending to about 30.

Stratus A layered cloud found at all atmospheric levels. Stratus is formless, sometimes thick, sometimes thin and wispy, with color ranging from gray to pearl white. Stratus forms when the atmosphere is stable. Precipitation associated with it is usually mist.

Sun dog Sometimes called mock sun, sun dogs are two bright spots of color at 22 degrees from and usually on either side of the sun at the same elevation. They are produced by the refraction of light passing through ice crystals.

Sun pillar Usually seen at sunrise or sunset, sun pillars are shafts of bright white light that stretch from the ground up into the sky. They are produced by the reflection of sunlight on flat faces of ice-crystal plates.

Sunrise/sunset The glorious colors often seen in clouds at sunrise and sunset are due to the scattering of different wavelengths of sunlight. The scattering is caused by dust particles and water vapor in the air.

Supercell A large cluster of cumulonimbus that develops into a very dangerous storm with a single violent updraft and, at times, a vortex of spinning air that can turn into a tornado.

Supercooled cloud droplets Liquid cloud droplets that can be observed at temperatures as low as minus 40 degrees F.

Supersaturated air A condition that occurs in the atmosphere when the relative humidity is greater than 100 percent.

Swelling cumulus Clouds in the cumulus family that swell to greater heights than fair-weather cumulus as an increase in surface heating gives them greater buoyancy.

Thunder The sound that occurs when the abrupt heating in the air from a lightning bolt causes a pressure wave that travels at the speed of sound.

Thunderstorm A storm produced by cumulonimbus clouds that produce lightning and thunder.

Tornado An intense, rotating column of air that extends from the base of a cumulonimbus supercell in the shape of a funnel and touches the ground.

Tropical depression An unsettled weather region that usually forms in the tropical Atlantic or Pacific. As it progresses eastward, the disturbance can intensify and develop a cyclonic pattern of winds up to 39 miles per hour.

Tropical storm A stormy region in the tropics with cyclonic winds up to 73 miles per hour.

Troposphere The layer of the atmosphere extending from the earth's surface to seven miles high.

Trough An elongated area of low atmospheric pressure.

Typhoon A hurricane that forms over the western Pacific Ocean.

Unstable air A condition in the atmosphere that exists when lifted air is warmer than the air around it.

Virga Moisture that evaporates as it falls from a cloud into a layer of drier air.

Wall cloud An area of rotating clouds that extends beneath a severe thunderstorm and from which a funnel cloud may appear.

Warm front A gently sloping boundary plane along which warm air overrides cold air.

Water vapor Water in a vapor, or gas, form.

Wave cyclone An extratropical cyclone that forms and moves along the polar front. The circulation of winds around the cyclone produces a wavelike deformation on the front.

Wind shear A rapid change of wind speed or wind direction over a short distance.

Index

Page numbers in *italics* refer to illustrations and captions.

ACKNOWLEDGMENTS

The invitation to write this book came as a complete surprise. It happened because Barbara J. Morgan, publisher and editorial director of Silver Lining Books, was searching the Internet for an author to write a book on clouds. When she came across my Web site, www.cloudman.com, she saw in it a possibility that ultimately became a reality.

Cloud lovers around the world have given me positive feedback on the brilliance of the cloud images and the sensitive organization of the site. Had it not been for the good work of Anthony and Janice Richardson of DoubleRichDesign, Barbara and I might never have made contact.

The magic touch of Richard Berenson, designer, is to be felt in every chapter, but particularly in the Cloud Portfolio. I am indebted to him, and wish to express gratitude.

I wish to thank the administrators of my college home, Linfield College in Oregon, for inviting me to continue to teach meteorology, my favorite subject, for 24 continuous years following my formal retirement in 1978. This has been an ongoing challenge, and it's a source of great satisfaction to rub shoulders with the younger generation. It has kept me alive.

Much is owed to my longtime friend J. Jack Borden, creator of the For Spacious Skies movement, whose vision, zeal, and dedication to the concept of Sky Awareness were so infectious that they became my own. I also am grateful to my cloud physics colleague Vincent J. Schaefer, who in 1971 invited me to collaborate with him in writing the *Peterson Field Guide to the Atmosphere* (1981). This book opened multiple doors of opportunity in my professional life.

My chapter on Luke Howard should serve only as a brief introduction to this man's extraordinary contributions to the study of clouds. Readers will find a more thorough treatment of Luke Howard's life and times in Richard Hamblyn's fine book, *The Invention of Clouds,* published by Farrar, Straus and Giroux, New York.

And last, but certainly not least, I thank my wife and beloved companion of 65 years and our five children and their families, who have patiently watched me grow in understanding. They have contributed more to the person I am than they can ever know.

ADDITIONAL PICTURE CREDITS

METRIC EQUIVALENTS CHART
Inches to Millimeters and Centimeters
MM=Millimeters   CM=Centimeters

Inches	MM	CM	Inches	CM	Inches	CM
1/8	3	0.3	9	22.9	30	76.2
1/4	6	0.6	10	25.4	31	78.7
3/8	10	1.0	11	27.9	32	81.3
1/2	13	1.3	12	30.5	33	83.8
5/8	16	1.6	13	33.0	34	86.4
3/4	19	1.9	14	35.6	35	88.9
7/8	22	2.2	15	38.1	36	91.4
1	25	2.5	16	40.6	37	94.0
1-1/4	32	3.2	17	43.2	38	96.5
1-1/2	38	3.8	18	45.7	39	99.1
1-3/4	44	4.4	19	48.3	40	101.6
2	51	5.1	20	50.8	41	104.1
2-1/2	64	6.4	21	53.3	42	106.7
3	76	7.6	22	55.9	43	109.2
3-1/2	89	8.9	23	58.4	44	111.8
4	102	10.2	24	61.0	45	114.3
4-1/2	114	11.4	25	63.5	46	116.8
5	127	12.7	26	66.0	47	119.4
6	152	15.2	27	68.6	48	121.9